Green to Gold

Green to Gold

How Smart Companies Use Environmental Strategy to Innovate, Create Value, and Build Competitive Advantage

Daniel C. Esty and Andrew S. Winston

Yale University Press

New Haven and London

Set in Adobe Garamond with Stone Sans Display by Westchester Book Services.

Printed in the United States of America.

Library of Congress Cataloging-in-Publication Data

Esty, Daniel C.
Green to gold : how smart companies use environmental strategy to innovate, create value, and build competitive advantage / Daniel C. Esty and Andrew S. Winston.
p. cm.
Includes bibliographical references and index.
ISBN–13: 978–0–300–11997–8 (cloth : alk. paper)
ISBN–10: 0–300–11997–6 (cloth : alk. paper)
1. Industrial management—Environmental aspects. 2. Corporations—Environmental aspects. 3. Business enterprises—Environmental aspects. I. Winston, Andrew S. II. Title.
HD30.255.E88 2006
658.4'083—dc22 2006022012

A catalogue record for this book is available from the British Library.

This book was printed with soy-based ink on acid-free recycled paper that contains 100% postconsumer fiber. The case was manufactured using acid-free recycled paper that contains postconsumer pulps.

The carbon dioxide emissions associated with the publication and distribution of this book have been offset by the purchase of carbon credits.

15 14 13 12 11 10

For our children—Sarah, Thomas, and Jonathan (the Estys), and Joshua and Jacob (the Winstons).

We hope you will be the beneficiaries of a future where businesses both profit and help to create a healthy and sustainable world.

Contents

** References are not noted in the text but are shown by page number and key words beginning after the appendices.*

Acknowledgments

We were granted remarkable access to top officials in dozens of companies. We are profoundly grateful for the many hours environmental professionals, factory managers, division heads, Board Members, COOs, and CEOs spent with us. We also interviewed people working closely with industry at major environmental groups. In total, we spoke to more than 300 people at more than 100 companies.

We can't thank individually all the executives who shared with us their environmental challenges and triumphs with impressive candor and humor. But we do want to single out a few people who really went beyond the call of duty and cleared a path for us. They arranged extensive company visits for us, gave us access to all levels of the company, ensured that we got into "the trenches," and often sat through multiple days of meetings listening to stories they knew already. In this regard, we thank Keith Miller and Kathy Reed at 3M, Shaye Hokinson at AMD, Chris Mottershead at BP, Pat Nathan at Dell, Dawn Rittenhouse and Paul Tebo at DuPont,

Steve Ramsey and Mark Stoler at GE, Kent Gawart and Kris Manos at Herman Miller, Thomas Bergmark at IKEA, Tim Mohin at Intel, Larry Rogero at FedEx Kinko's, Bob Langert at McDonald's, Albin Kaelin at Rohner Textil, Mark Weintraub at Shell, Terry Kellogg at Timberland, and Clive Butler at Unilever.

Two authors working alone could never put together a book like this. We have many people to thank.

We owe a great debt to the Yale community. A dedicated team of research assistants from the Yale School of Forestry and Environmental Studies, the Yale School of Management, and the Yale Law School worked tirelessly to provide data, analysis, and in-depth research to help back up or redirect our theories and ideas. We want to particularly acknowledge the contributions of Pat Burtis, Pamela Carter, Genevieve Essig, Jordanna Fish, Cassie Flynn, Jennifer Frankel-Reed, Rachel Goldwasser, Kaitlin Gregg, Ann Grodnik, Lauren Hallett, Laura Hess, Andrew Korn, Cho Yi Kwan, Emily Levin, Jessica Marsden, Tiffany Potter, Marni Rappaport, Kara Rogers, Elena Savostianova, Manuel Somoza, Grayson Walker, Austin Whitman, and Rachel Wilson.

Melissa Goodall and Christine Kim at the Yale Center for Environmental Law and Policy helped shepherd this project through its many phases and always stayed calm under fire. Special thanks to Marge Camera at the Yale Law School, who was the first person to read this book cover to cover while making our countless edits a reality in digital form.

We also want to thank our agent, Rafe Sagalyn, for ensuring that we stayed on track and our publicist Barbara Henricks for guiding us through the media jungle. Special thanks as well to our editorial advisor, Howard Means, for helping us set the right tone throughout the book. We are deeply grateful to our Yale editor Mike O'Malley and the rest of the team at Yale University Press and Westchester Books, including Steve Colca, Jessie Hunnicutt, Debbie Masi, Cher Paul, Liz Pelton, and Mary Valencia for bringing our vision into reality.

The quality of the book was greatly enhanced by the perspectives of a group of executives, scholars, and advisors who provided intellectual support from the onset. Special appreciation goes to Mike Porter from the Harvard Business School who, from the first days of this project, has helped us think through the challenges of raising environmental strategy to a higher level. We've benefited from comments and suggestions from many others, including Antony Burgmans, Chantal-Line Carpentier, Betrand Collumb, Daniel Gagnier, Brad Gentry, Diana Glassman, Hank Habicht, Chad Holliday, Robert Jacobson, Harri Kalimo, Nat Keohane, Frank Loy, Pat McCullough, Raymond Necci, Mads Ovlisen, Robert Repetto, Jeff Seabright, Jeff Sonnenfeld, David Vogel, Dennis Welch, Richard Wells, and Tensie Whelan. Special thanks to Gordon Binder, Marian Chertow, Bill Ellis, Larry Linden, and Jan Winston (Andrew's father), each of whom read our draft with an incredible eye for detail and helped us refine how we laid out the concepts and ideas in the book. A special acknowledgement to Matt Blumberg who provided support ranging from the concrete, such as office space in New York, to the intellectual, such as perspectives on how to expand the appeal of the text to small businesses and how to market the book.

We also wish to express deep gratitude for the generous support, both financial and intellectual, we received from a number of foundations and their environmental leaders: the Johnson Foundation and Jesse Johnson; the Surdna Foundation and Ed Skloot and Hooper Brooks; the Overbrook Foundation and Daniel Katz; and Fletcher Asset Management, Inc., and Alphonse Fletcher, Jr. And special thanks to the Betsy and Jesse Fink Foundation for supporting our outreach efforts—and to Jesse, who helped to shape the intellectual agenda behind this project from its earliest days.

Finally, to our very patient wives, Elizabeth and Christine: Thank you for your support in the most fundamental ways—and for listening sympathetically to our musings and theories as well as putting up with the odd work hours of a writer.

Preface

In the run-up to the 1992 Earth Summit in Rio de Janeiro, business leaders began to focus on environmental issues as never before. Organized by Swiss billionaire Stephan Schmidheiny, fifty leading companies formed a Business Council for Sustainable Development. At the same time, Schmidheiny and his colleagues wrote a book, *Changing Course*, and launched the concept of eco-efficiency, emphasizing the potential economic gains from reducing pollution and better managing natural resources. Hundreds of CEOs attended the Rio convocation, and thousands of others were inspired to ask how their companies could become better environmental stewards.

Dan Esty participated in the Earth Summit as a U.S. Environmental Protection Agency official. The U.N. Conference on Environment and Development, as it was formally called, generated unprecedented focus on threats to the natural world from climate change to biodiversity loss. Discussions centered on how all parts of society could act together to address the problems. Hope ran high. In the immediate aftermath, a range of companies promised big action to reduce their environmental impacts.

In 2002, many of the same players reassembled, this time in Johannesburg, South Africa, for the World Summit on Sustainable Development. Dan attended again, this time as a representative of Yale University. But something had changed. Little progress had been achieved on most of the major issues identified at the Rio conclave a decade earlier. The environmental, or "green," call to arms of ten years earlier seemed to have run out of steam. Many companies had adopted environmental policies and even taken "beyond compliance" approaches to pollution control over the previous decade. But a more cynical view about the business world's environmental agenda and prospects for improvement had taken root. What happened? Surely the job was not done. To the contrary, the environmental movement also seemed to be losing momentum. Where was that earlier energy and enthusiasm? What might be done to reinvigorate corporate environmental strategy?

These questions and others were on Dan's mind when Andrew Winston arrived at the Yale School of Forestry and Environmental Studies in 2002. After a ten-year career in marketing, business development, and strategy, Andrew had decided to focus his energies on how companies address environmental issues. With a passion both for developing winning business strategies and for protecting the planet, Andrew wanted to bridge the gap between two worlds that usually moved in separate orbits.

To make sense of the corporate environmental strategy situation, we launched a review of what companies had been doing on this front for the last decade. We drew on Dan's fifteen-year career working with companies to improve their strategies and on his class, Corporate Environmental Strategy and Competitive Advantage, at Yale and at INSEAD (the European business school in Fontainebleau, France). Our first step was to examine closely the canon of green business—the major books, articles, and case studies that had discussed the business–environment interface.

We were shocked by what we found. Most of the literature focused on "win-win" outcomes. Indeed, many of the books and articles had a cheerleading tone. Over 95 percent of the stories and examples talked only about the benefits of environmental thinking—reducing environmental impact and saving money. Surely, we thought, these initiatives can't always be successful. No business strategy works *all* the time. Could this one-sided perspective and lack of analytic rigor be one of the reasons a broader business commitment to environmental action had not really taken hold? Were average business people skeptical of the unremittingly positive claims of green gurus? Where was the business-like edge and hard-hitting advice?

To see what's really happening at the interface of business and the environment, we've spent the last four years talking to hundreds of people at companies, industry associations, and environmental groups—and poring over the data. We haven't shied away from the success stories—they tell us a great deal. However, we've also studied what didn't work and why initiatives that often looked good on paper have failed in fact. Companies embarking on environmental efforts of their own shouldn't have to repeat the mistakes of the past.

The result, we hope, is a thorough review of what works—and what doesn't—when companies fold environmental thinking into business strategy. As we'll show, the stakes are high—environmental issues are real and pressing. And more and more "stakeholders" care deeply about how companies act and are not afraid to pressure them to do more. But the potential rewards are great too. With *Green to Gold*, we hope to blaze a trail for managers and executives toward stronger businesses *and* a healthier planet.

Daniel C. Esty
Andrew S. Winston
March 2006
New Haven, Connecticut
New York, New York

Introduction The Environmental Lens

SONY'S VERY EXPENSIVE CHRISTMAS

In the weeks before Christmas 2001, the Sony Corporation faced a nightmare. The Dutch government was blocking Sony's entire European shipment of PlayStation game systems. More than 1.3 million boxes were sitting in a warehouse instead of flying off store shelves. Was this a trade war or an embargo against violent video games? Sony executives probably wished it were something that easy to fix.

So why was Sony at risk of missing the critical holiday rush? Because a small, but legally unacceptable, amount of the toxic element cadmium was found in the cables of the game controls. Sony rushed in replacements to swap out the tainted wires. It also tried to track down the source of the problem—an eighteen-month search that included inspecting over 6,000 factories and resulted in a new supplier management system. The total cost of this "little" environmental problem: over $130 million.

Sony executives refer to their PlayStation disaster as the "Cadmium Crisis." They've vowed never again to be caught unaware of environmental risks. In fact, they're now much more familiar with their own operations as a result of hunting down the problem.

So what can we learn from all this? Did an environmental ogre get what it deserved? Hardly. Sony has been a business powerhouse for years, and despite a few hiccups, the company is also generally perceived as an environmental leader. Nothing, in fact, foreshadowed the PlayStation stumble, yet it happened. Why? From Sony's difficult experience, we draw three lessons:

- Even the best companies can be surprised by environmental issues.
- The environment is not a fringe issue—it can cost businesses real money.
- Real benefits can come from seeing things in a new light.

BP AND "LOOKING FOR CARBON"

While Sony's game systems sat in a warehouse, another very large but very different company was counting the money it saved when it sharpened its environmental focus and started looking at its business in a different way.

BP's chief executive, Lord John Browne, committed the company to reducing its emissions of the greenhouse gases that contribute to global warming, especially carbon dioxide. Browne told all of BP's business units to find ways to produce less of these gases. And they did. After three years of what insiders call "looking for carbon," BP discovered numerous ways to cut emissions, improve efficiency, and save money. A lot of money.

The initial process changes cost BP about $20 million but saved the company an impressive $650 million over those first few years. As of 2006, the savings topped $1.5 billion. In a low-key British way, BP executives told us that they were floored by the outcome. Nobody had dared imagine such an absurdly high return on investment. As Browne has said, "We set out to do good . . . and we ended up doing well."

Was BP radically inefficient before this program? Far from it. The company had just never looked at its operations with an eye toward reducing greenhouse gas emissions. Once it did, innovation flourished, all to the benefit of the bottom line.

Looking at all the ways environmental issues affect a business can frame thinking and strategy in a new way. By examining their business through an environmental "lens," managers can avoid expensive problems and create substantial value. Thus, we add a fourth lesson to the three we drew from Sony's experience—and this is the fundamental one:

Smart companies seize competitive advantage through strategic management of environmental challenges.

BP and Sony learned what some companies already knew: The Business world and the natural world are inextricably linked. Our economy and society depend on natural resources. To oversimplify, every product known to man came from something mined or grown. The book you're reading was once a tree; the ink these words are printed in began life as soybeans. The environment provides critical support to our economic system—not financial capital, but natural capital. And the evidence is growing that we're systematically undermining our asset base and weakening some of our vital support systems.

In other words, an environmental lens is not just a nice strategy tool or a feel-good digression from the real work of a company. It's an essential element of business strategy in the modern world. It provides a way for businesses to contend with the real problems of pollution and natural resource management. Mismanaging these issues can drain value out of a company quickly—and damage brand reputations built up over decades of careful cultivation. That's why leading companies have learned to manage environmental risks and costs as closely as they do other risks and costs. In doing so, they reduce the risk to the whole enterprise.

But the upside is equally important. In the chapters that follow, we'll explore how leading companies are layering environmental (often called "green") factors into their corporate strategies—spurring innovation, creating value, and building competitive advantage. These leaders see their businesses in fresh ways. They create new products to meet environmental needs. As they look up and down the value chain, they keep environmental impacts firmly in mind. They know that working to protect the planet also protects their own companies—by safeguarding their assets, inspiring current employees, and attracting valuable new "knowledge workers" looking for more than a paycheck.

In *Green to Gold*, we take you inside leading companies, across industries, and around the world. We show you the real costs, hard choices, and trade-offs companies face when they make environmental thinking part of their core business strategy. Pundits who dismiss the natural world as an issue—or commentators on the other "side" who underestimate the difficulties businesses face in executing environmental strategies—do neither the business world nor the planet any favors.

By systematically analyzing the experiences of dozens of companies, we've been able to extract the key strategies, tactics, and tools that are needed to establish an environmentally based competitive advantage. In a marketplace where other points of competitive differentiation, such as capital or labor costs, are flattening, the environmental advantage looms larger as a decisive element of business strategy. Indeed, no company can afford to ignore green issues. Those who manage them with skill will build stronger, more profitable, longer-lasting businesses—and a healthier, more livable planet.

Part One **Preparing for a New World**

In the first few chapters, we lay out the context for this book, highlighting how environmental challenges have become an important part of the business landscape. In Chapter 1, we introduce the "Green Wave" sweeping the business world, and we present the logic for making environmental thinking a core part of strategy. We also spell out some of the "megaforces," like globalization, that give the new environmental imperative greater prominence. Finally, we provide an overview of how we conducted our research and picked the companies that are the focus of this book.

Chapters 2 and 3 introduce the new pressures—both natural and human—coming to bear on companies. These forces make attention to environmental strategy essential for business success. We start in Chapter 2 by highlighting the environmental problems facing humanity and every company, ranging from global warming to water shortages. For each issue in our environmental primer, we offer a crisp summary of the problem, a review of the possible range of effects, and an analysis of how the problem might affect business.

In Chapter 3, we review the growing array of environment-oriented "players" on the field of business. We map 20 different categories of stakeholders from traditional government regulators to powerful nongovernmental organizations (NGOs) to increasingly environmentally focused banks. We highlight the questions these groups are asking about how companies operate.

In brief, this section explains how and why the environment has emerged as a critical strategic issue for companies of all sizes. It sets the stage for our tour of the critical elements of corporate environmental strategy. And it shows how careful thinking about the environment can provide a new basis for competitive advantage.

Chapter 1 Eco-Advantage

Washington, D.C.: General Electric CEO Jeff Immelt announces a new initiative, "ecomagination," committing the mega-manufacturer to double its investment in environmental products—everything from energy-saving lightbulbs to industrial-sized water purification systems and more efficient jet engines. Backed by a multi-million-dollar ad campaign, Immelt positions GE as the cure for many of the world's environmental ills.

Bentonville, Arkansas: In a speech to shareholders, Wal-Mart CEO Lee Scott lays out his definition of "Twenty First Century Leadership." At the core of his new manifesto are commitments to improve the company's environmental performance. Wal-Mart will cut energy use by 30 percent, aim to use 100 percent renewable energy (from sources like wind farms and solar panels), and double the fuel efficiency of its massive shipping fleet. In total, the company will invest $500 million annually in these energy programs. Moreover, in a move with potentially seismic ripples, Wal-Mart will "ask" suppliers to create more environmentally friendly products: some of the fish Wal-Mart sells will have to come from sustainable fisheries, and the clothing suppliers will use materials like organic cotton. "We believe that these initiatives will make us *a more competitive and innovative company*," Scott emphasizes.

By either market cap or sales, GE and Wal-Mart are two of the biggest companies in human history. Neither company springs readily to mind when you say the word "green." But these are not isolated stories. Companies as diverse as Goldman Sachs and Tiffany have also announced environmental initiatives. As the *Washington Post* observed, GE's move was "the most dramatic example yet of a green revolution that is quietly transforming global business."

What's going on? Why are the world's biggest, toughest, most profit-seeking companies talking about the environment now? Simply put, because they have to. The forces coming to bear on companies are real and growing. Almost without exception, industry groups are facing an unavoidable new array of environmentally driven issues. Like any revolution, this new "Green Wave" presents an unprecedented challenge to business as usual.

NEW PRESSURES

Behind the Green Wave lie two interlocking sources of pressure. First, the limits of the natural world could constrain business operations, realign markets, and perhaps even threaten the planet's well-being. Second, companies face a growing spectrum of stakeholders who are concerned about the environment.

Global warming, water scarcity, extinction of species (or loss of "biodiversity"), growing signs of toxic chemicals in humans and animals—these issues and many others increasingly affect how companies and society function. Those who best meet and find solutions to these challenges will lead the competitive pack.

The science, we stress, is *not* black and white on all these issues. Some problems, like ozone layer depletion or water shortages, are fairly straightforward. The trends are plainly visible. On other issues—climate change most notably—uncertainties persist, but the evidence is clear enough and the scientific consensus strong enough to warrant action.

THE EVOLVING CHALLENGE

Environmental worries used to center on "limits to growth" and the prospect of running out of key natural resources like oil and industrial metals. These concerns have often been overblown. A second line of concerns focused on pollution and has proven to be more enduring. We now know that humans *can* overwhelm the capacity of nature—from local waterways to the global atmosphere—to absorb pollutants or to provide the essential "ecosystem services" we need, such as fresh water, breathable air, a stable climate, and productive land.

A broad-based set of players now insists on attention to these issues. Government, the traditional superpower of influence on corporate behavior, has not gone away. Far from it. Regulators worldwide no longer turn a blind eye to pollution. Citizens simply won't allow it. Across all societies, we see serious efforts to control emissions and make polluters pay for the harm they cause.

Other actors, however, now play prominent environmental roles on the business stage. NGOs, customers, and employees increasingly ask pointed questions and call for action on a spectrum of issues. To give just one example, HP says that in 2004, $6 billion of new business depended in part on answers to customer questions about the company's environmental and social performance—up 660 percent from 2002. These demands reshape markets, create new business risks, and generate opportunities for those prepared to respond.

The breaking news is the arrival of a new set of stakeholders on the environmental scene, including banks and insurance companies. When the financial services industry—which focuses like a laser on return on investment—starts worrying about the environment, you know something big is happening. Wall Street stalwart Goldman Sachs announced that it would "promote activities that protect forests and guard against climate change." The company said it would "encourage" its clients to act greener and pledged $1 billion for investment in alternative energy, having already bought a company that builds wind farms. Even for Goldman, a billion dollars is not a token gesture. JPMorgan, Citigroup, and other big names have made similar commitments—and they have signed on to the "Equator Principles," which require environmental assessments of major loans.

For a painful example of how this one-two punch of natural forces and new stakeholders can slam a company, just ask Coca-Cola's two most recent ex-CEOs, Doug Ivestor and Doug Daft. Within the past five years alone, the world's largest soft-drink manufacturer faced angry protests in India over its water consumption, came under pressure to stop using refrigerants that hurt the ozone layer, and withdrew its flagship bottled water Dasani from the British market after the supposedly purified drink failed European Union quality tests. Today, the company has a vice president dedicated to water and environment issues, Jeff Seabright, and a new CEO, Neville Isdell, who works closely with the company's Environmental Advisory Board, on which one of us (Dan) serves.

Environmental missteps can create public relations nightmares, destroy markets and careers, and knock billions off the value of a company. Companies that do not add environmental thinking to their strategy arsenal risk missing upside opportunities in markets that are increasingly shaped by environmental factors.

THE BUSINESS CASE FOR ENVIRONMENTAL THINKING

We see three basic reasons for adding the environmental lens to core strategy: the potential for upside benefits, the management of downside risks, and a values-based concern for environmental stewardship.

The Upside Benefits

Nobody, not even market-savvy Toyota, could have predicted the success of its hybrid gas-electric Prius. Given the poor track record of electric vehicles, this leap of faith was anything but a clear path to profit. Yet Toyota executives saw potential value down the road, and they could not have been more correct. After a decade-long research push, the Prius was named *Motor Trend*'s Car of the Year in 2004, by which time customers were waiting six months to get their hybrid cars. While Detroit was nearing bankruptcy, laying off tens of thousands of workers, and offering "employee discounts" to everybody, Toyota was raising prices, expanding production, collecting record profits of $11.8 billion in 2006, and closing in on the title of world's largest automaker.

Toyota's green focus is no accident. In the early 1990s, when Toyota wanted to design the 21st-century car, it made the environment a major theme, ahead of all the selling points that automakers traditionally used: size, speed, performance, or even ability to attract beautiful girls or hunky guys. Smart move.

Similarly, BP has rebranded itself as an energy company, preparing to move "beyond petroleum" and investing in renewable energy. These companies have figured out that it's better to remake your marketplace and eat your own lunch before someone else does.

Our research suggests that companies using the environmental lens are generally more innovative and entrepreneurial than their competitors. They see emerging issues ahead of the pack. They are better prepared to handle the unpredictable forces that buffet markets. And they are better at finding new opportunities to help customers lower their costs and environmental burden. By remaking their products and services to respond to customer needs, they drive revenue growth and increase customer loyalty.

The "gold" that smart companies mine from being green includes higher revenues, lower operational costs, and even lower lending rates from banks that see reduced risk in companies with carefully constructed environmental management systems. They also reap soft benefits, from a more innovative culture to enhanced "intangible" value, credibility, and brand trust.

Scholars and pundits have noted that businesses now face a world where traditional elements of competitive advantage, such as access to cheaper raw materials and lower cost of capital, have been commoditized and whittled away. On this altered playing field, going green offers a vital new path to innovation and to creating enduring value and competitive advantage. Nike executive Phil Berry puts it simply: "We have two maxims. Number 1: It is our nature to innovate. Number 2: Do the right thing. But everything we do around sustainability is really about number one—it's about innovation."

The Downside Risks

Inside oil giant Shell, executives use the acronym TINA—There Is No Alternative—to explain why they do some things. To them, thinking about how climate change affects their business or caring how stakeholders feel about the company is no longer optional. It's just a fact of life. Even through well-publicized problems with local communities and governments in places like Nigeria, Shell has continued to hone its stakeholder relations skills. The company spends millions of dollars working with the people living around key oil and gas projects such as the massive Athabasca Oil Sands in Alberta, Canada.

As head of Shell's famed scenarios group, Albert Bressand helped the executive team think about what could hurt the company in the long term. As he told us, "We are a prisoner of the market. . . . there are people who can remove our license to operate."

The idea behind license to operate is simple: society at large *allows* companies to exist and gives them a certain leeway. If your company oversteps the bounds, societal reactions can be harsh and, in severe cases, destroy the company. Former partners of Arthur Andersen learned that lesson at great cost when the accounting giant vanished in the wake of the Enron scandal. Or remember the case of chemical industry leader Union Carbide? After the company's 1984 disaster in Bhopal, India, which killed over 3,000 people, Union Carbide's future fell apart until finally it was swallowed by Dow.

More pointedly, society's expectations about company behavior are changing. A company that abuses the local environment can find it impossible to get permits to expand operations. Regulators, politicians, and local communities raise fewer barriers for good neighbors.

Heavy industries are especially aware of this social license issue, but others feel the heat as well. After years of unfettered expansion, Wal-Mart has come under fire from protestors who contend that the company's stores increase sprawl, destroy wetlands, and threaten water supplies. In some communities, regulators have joined the chorus and begun to impinge on the retail giant's expansion plans. In internal meetings, Lee Scott told Wal-Mart executives that their sustainability efforts would help protect the company's "license to grow."

Environmental challenges can seem like a series of small holes in a water main, slowly draining value from the enterprise. Or they can appear suddenly as major cracks in a dam and threaten the entire business. Maybe the problem is unexpected costs for pollution control or a cleanup for which nobody budgeted. Maybe it's a very public disaster like the *Exxon Valdez*. Sometimes, too, the downside of mismanaging these issues can get very personal. Executives who preside over the mishandling of toxic waste, for example, can face jail time.

Efforts to cut waste and reduce resource use, often called "eco-efficiency," can save money that drops almost immediately to the bottom line. Redesign a process to use less energy, and you'll lower your exposure to volatile oil and gas prices. Redesign your product so it doesn't have toxic substances and you'll cut regulatory burdens—and perhaps avoid a value-destroying incident down the road. These efforts lower business risk while protecting the gold—reliable cash flows, brand value, and customer loyalty, for example—that companies have painstakingly collected over time.

Smart companies get ahead of the Green Wave and lower both financial and operational risk. Their environmental strategies provide added degrees of freedom to operate, profit, and grow.

The Right Thing to Do

Repeatedly during our research we asked executives why their companies launched environmental initiatives, some of which cost significant money up front and had uncertain paybacks. More often than you might imagine—and far more often than we first expected—they said that it was the right thing to do.

Is the case for thinking and acting environmentally based on values? Not primarily. At least that's not what we heard from the executives we interviewed. For most of them, the moral argument was not a separate imperative. It was deeply intertwined with business needs. Building a company with recognized values has become a point of competitive advantage, whether you have 2 employees or 200,000. Doing the right thing attracts the best people, enhances brand value, and builds trust with customers and other stakeholders. In fact, it's hard to conceive of a business asset more central to long-term success than trust among stakeholders—or one that is more easily lost. As investing legend Warren Buffett once said, "It takes twenty years to build a reputation and five minutes to ruin it. If you think about that, you'll do things differently."

Even those who agree with Nobel Prize–winning economist Milton Friedman that the main "social responsibility of business is to increase its profits" can't ignore the growing ranks who believe that companies have an obligation to do more. The logic of corporate environmental stewardship need not stem from a personal belief that caring for the natural world is the right thing to do. If critical stakeholders believe the environment matters, then it's the right thing to do *for your business*.

Environmental leaders see their businesses through an environmental lens, finding opportunities to cut costs, reduce risk, drive revenues, and enhance intangible value. They build deeper connections with customers, employees, and other stakeholders. Their strategies reveal a new kind of sustained competitive advantage that we call Eco-Advantage.

MAGNIFYING FORCES

The Green Wave, with its threats and opportunities, rises within a business landscape already in the throes of radical change. Companies face a number of mega-trends that interact with the effects of the Green Wave, accelerating change and magnifying its impact and scope.

Globalization and Localization

As author Thomas Friedman describes it, outsourcing is just the tip of the iceberg. The "flattening" of the global markets for goods and services will disrupt nearly all industries. The continued rise of both China and India seems likely to have a profound effect on businesses across the world, especially in North America and Europe.

Economic integration and trade liberalization intensify competition. Globalization creates opportunities for many, but fundamentally rewards scale. Size, however, creates suspicion of excess power. Large enterprises come under extra scrutiny for their business practices, including environmental impacts.

Simultaneously, the world is fragmenting, with niche markets demanding tailored products and services. For example, in many of its restaurants in India, McDonald's serves curry, not its brand-defining hamburgers. Operating in ways that respond to localized needs and preferences is becoming essential. The scale of environmental issues, ranging from entirely local to inescapably global, only adds to the complexity of this already daunting management challenge.

Insecurity

The security tensions that have swept through the United States and much of the rest of the world in the wake of 9/11 are changing public attitudes and the political landscape. Beyond fears of terrorism, there's a growing public distress over reliance on oil from the Middle East. Increasingly, people express a willingness to pay a premium for fuel supplies closer to home, including alternative energy sources such as wind or solar power. Thus, the energy future looks very different from the past, with profound environmental consequences.

Government ↓, Business ↑

While the era of big government is over, public expectations about the role of business in meeting society's needs are rising. Many would say the regulatory system in America is becoming less burdensome, but in Europe regulations seem to be weighing more heavily on business. What's common throughout the world is that companies are being asked to do more voluntarily, not just for the environment but for a constellation of social issues including poverty alleviation, education, and health care. The debate about the appropriateness of these expectations rages on, but the trend is clear. An intensified focus on corporate social responsibility, or CSR, is here to stay.

Big business faces even more elevated expectations. Multinational companies, with their global reach and ubiquitous effects, are held to higher standards than are smaller companies. To those whom much is given, much is expected. When operating on foreign soil, companies must expect especially intense scrutiny. One telling example: While Coca-Cola faces ongoing protests in India over its water use at a plant in Kerala, the Indian-owned Kingfisher brewery down the road, which uses far more water, draws no political ire.

Rise of the Middle Class in Emerging Economies

Let's look at just one statistic: The number of cars in China and India is predicted to rise from about 17 million today to 1.1 *billion* (yes, billion) by 2050. The addition of hundreds of millions of working, middle-class people in the developing world, all seeking a Western quality of life, will shake up nearly every industry. For those who are prepared, this new market offers considerable opportunity. But the same growth in consumption threatens to destroy natural resources and inflict pollution on the planet on an unprecedented scale.

Continuing Pressure from Poverty

Despite substantial middle-class growth, particularly in Asia, many parts of the developing world continue to struggle with chronic poverty. Expanding populations strain limited resource bases. Poverty forces short-term thinking that translates into environmental degradation. People cut down trees for fuel, for example, without regard to the soil erosion and other negative consequences that follow.

So while the rise of a consuming middle class creates one set of environmental threats, persistent poverty represents an equally serious social and ecological challenge that the business world avoids at its own risk. As a top executive from ABN AMRO Real—the Brazilian subsidiary of the Dutch bank and a company on the front lines in the developing world—observed, "Companies cannot succeed in societies that fail."

Better Science and Technology

The science underlying many environmental problems is getting clearer. But the downside of clarity is an increased obligation to respond. "Biomonitoring" and more sensitive measurement tools can now identify—at trace levels—virtually every chemical or emission found in the environment, whether in a polar bear in the Arctic Circle or in the breast milk of a woman in Ohio. Even exposures at the parts-per-billion level can trigger calls for action from an expanding army of advocacy groups. On the positive side, new developments such as nanotechnologies could provide solutions to environmental ills—and create market opportunities for entrepreneurial businesses.

Transparency and Accountability

"The web changes everything" seems like yesterday's news, but the ripples of the Digital Age continue to move through our economy and society. The famous "Moore's Law" predicts that the density of transistors on a microchip will double every 18 months. This trend has held for 40 years and relentlessly drives computing power up and the costs of digital technology down. For billions of people, an endless variety of information—and perhaps a degree of disinformation—is just a click away.

The unprecedented level of transparency provided by the Internet is transforming the business world. With bloggers everywhere, including inside companies, anything that goes wrong anywhere in your operations—or in your suppliers' operations—can hit the web nearly instantaneously. As the *New York Times* put it, the Internet "has given the angry voices a more public outlet. The blogosphere is rife with postings castigating Coca-Cola, Wal-Mart, and other big companies, citing everything from water consumption to unfair labor practices and dangerous smokestack emissions." And this is not just idle chatter among uber-geeks. The conversations that bloggers start can move from cyberspace to Main Street in the blink of an eye.

In a world of rising transparency and low-cost information, *who* is responsible for *what* becomes increasingly clear. As pollution and toxic chemicals become easier to track back to their sources, we will know which companies created them, shipped them, used them, and disposed of them. No question about it: *Full* accountability is the emerging norm.

WHO SHOULD CARE

For some enterprises, a new green perspective will be transformative, leading to fresh thinking, new markets, profitability gains, and increased value. For others, the environmental lens may emerge more gradually and modestly, as another critical *element* of corporate strategy. With time, these companies may find long-term, sustained advantage, but not dramatic immediate gains, from being green. For the big, heavy industries, the gains are closer to being assured. But smaller and "cleaner" companies will find surprising benefits as well.

In today's world, no company, big or small, operating locally or globally, in manufacturing or services, can afford to ignore environmental issues. Of course, the opportunities and risks posed by the Green Wave vary by company and by industry. Context matters in the push for Eco-Advantage. No single strategy or tool will work in all companies or all circumstances. But Green Wave dynamics have become a fact of business life for nearly every organization. Companies that dive beneath the wave, submerging themselves in the hope that it will pass, will be disappointed by its enduring presence and pounding tenacity.

Why Small Businesses Should Care

What about small businesses? Can they sit this one out? In a word, no. Here are five reasons why.

- Laws that once applied only to big business are encroaching on smaller enterprises. Even bakeries and gas stations must now comply with clean air regulations.
- Going after the consumption choices of individuals remains difficult politically, but advocacy groups have no problem demanding that small businesses curb their impacts. So while personal cars may not come under NGO attack, the emissions from taxi fleets or delivery services make a relatively attractive target.
- The Information Age is reducing the costs of pursuing smaller-scale actors. New sensors, information systems, and communications technologies make tracking pollution and monitoring regulatory compliance cheaper every day. Even tiny enterprises now find it hard to fly under the radar.
- Large customers are putting pressure on small-business suppliers to comply with environmental standards. One little New York–based software developer we know found itself answering tough questions posed by a Tokyo-based telecom company with an aggressive auditing program for its supply chain. To stay on the list of preferred suppliers, the company had to implement an Environmental Management System—much more than a company its size would normally do.
- Small companies can be more nimble than their larger competitors. Entrepreneurial businesses can move quickly to take advantage of changing circumstances or meet niche demands. Q Collection, a "sustainable" home furnishings company, produces couches, tables, and chairs without toxic dyes and with wood sourced entirely from sustainably-managed forests. The furniture is priced at the high end of the market, but the company has found a customer base of interior designers who want the natural option. And Hawaii-based Kona Blue has launched an environmentally friendly fish farm to meet the growing demand for fish raised free of hormones and antibiotics.

WHAT DOES A COMPANY SEEKING ECO-ADVANTAGE LOOK LIKE?

We studied dozens of companies during our four years of research. A few have not evolved in their thinking since the 1970s. They are still grousing about legislation and complying with it grudgingly.

WHO SHOULD CARE THE MOST?

While we think this book is useful for anybody interested in a healthy environment and a healthy business community, clearly some companies need to worry about these issues more than others. And some sectors are poised for greater upside potential. We see growing risks and rewards for companies with:

- **High brand exposure.** Companies with substantial goodwill and intangible value (including Coca-Cola, Procter & Gamble, and McDonald's) face special challenges.
- **Big environmental impact.** Those in extractive industries or heavy manufacturing (BP, Exxon, Alcoa, and LaFarge, for example) must expect growing scrutiny.
- **Natural resource dependence.** Companies that sell fish, food, and forest products (such as Cargill, Nestlé, and International Paper) are likely to be on the front lines as society faces very real natural limits.
- **Current exposure to regulations.** Environmental strategy questions play a particularly important role for those handling hazardous materials (DuPont) or operating in heavily regulated industries like utilities (AEP).
- **Increasing potential for regulation.** Automakers and electronics producers (like Ford and Intel) are facing new challenges with European "takeback" laws that require manufacturers to handle the disposal of their products *after* their customers are done with them.
- **Competitive markets for talent.** Companies in the service sector and the "new economy" (such as Citigroup, Intel, or Microsoft)—where primary assets can walk out the door if they are displeased with the company's values—must stay on top of environmental issues.
- **Low market power.** Companies that rely on big customers that may start asking questions about environmental performance (most small-to-medium B2B companies) may be forced to raise their game. At the same time, those in highly competitive industries (such as small waste-handling businesses) will be hard pressed to step out in front of the competitors with initiatives that add costs or may not pay off for a long time.
- **Established environmental reputations.** Those with problematic histories should expect extra scrutiny. Companies with good track records will get more leeway—and may benefit from goodwill in the marketplace.

Others have begun to see the business opportunities in going "beyond compliance." A few have embarked on bold new initiatives to provide solutions to the world's environmental ills—like GE's plan to sell renewable energy, efficient power generation, water purification, and much more.

The companies who "get" the interface between environmentalism and business—the ones that are on their way to reducing their environmental impacts, or "footprints," while generating significant profits and sustained Eco-Advantage—have no single profile. They range from global conglomerates to niche textile makers. However, we found certain patterns. The leading-edge companies go beyond the basics of complying with the law, cutting waste, and operating efficiently. They fold environmental considerations into all aspects of their operations. Specifically, they:

- design innovative products to help customers with their environmental problems or even create new eco-defined market spaces;
- push their suppliers to be better environmental stewards or even select them on that basis;
- collect data to track their performance and establish metrics to gauge their progress;
- partner with NGOs and other stakeholders to learn about and find innovative solutions to environmental problems;
- build an Eco-Advantage culture through ambitious goal-setting, incentives, training, and tools to engage all employees in the vision.

For the top-tier companies, environmental management started out as something they had to do. But that's no longer the case. They've evolved to the point where environmental *management* is second nature and their focus is now on mining the gold in environmental *strategy*.

ENVIRONMENTAL STRATEGY, SUSTAINABILITY, AND CORPORATE SOCIAL RESPONSIBILITY

Companies find many ways to talk about how they handle environmental and social issues. Some focus on "triple bottom line" performance or sustainability. Others frame their work in terms of corporate social responsibility, stewardship, citizenship, or environment, health, and safety. Any of these approaches can serve to galvanize action and create Eco-Advantage. The key lies in execution—including environment and social issues in business operations. But each company needs to find the language and organizational structures that work within its own culture.

At the operational level, managing sustainability issues, no matter what the company calls them, works best with a defined focus. Thinking about environmental challenges alongside social issues such as health care, poverty alleviation, or how to serve the "bottom of the pyramid"—the untapped market of the world's poorest people—quickly becomes daunting. Our research suggests that the skills needed to manage environmental issues and social concerns are quite distinct. For example, what's required to ensure that a company complies with air-pollution permits, say, will have little similarity to what's needed to develop a strong employee wellness program.

Moreover, the environmental agenda has a concreteness that's often lacking on the social side. Obligations under the law are generally much clearer in the environmental realm, as are the opportunities for gaining a competitive advantage while doing the right thing. This is not to say that social issues are unimportant. Indeed, some are moral imperatives. As Professor David Vogel of Haas Business School has demonstrated, however, the *business case* for taking up the social agenda is much harder to establish. For all of these reasons, we focus on defining the strategies and tools companies can use to take advantage of *environmental* opportunities.

WHY SOCIAL ISSUES CAN'T BE IGNORED

While environmental and social issues pose different kinds of challenges, they both connect to corporate reputation. Any company that thinks it can cover shortcomings in social performance with strong environmental results is kidding itself. Wal-Mart, for example, has recently started to work on a range of issues from renewable energy to sustainable fishing to its impact on land use. But it shouldn't expect to win any prizes for corporate responsibility while falling short on basic social issues such as wages, health care, and labor relations.

ECO-ADVANTAGE IS NOT EASY

We wish we could say that finding Eco-Advantage will be easy. But like excellence in any form, you have to work for it. We know this runs contrary to the message in many of the books and articles about "green business." Ever since a few leaders like 3M demonstrated the payoffs of eco-efficiency, going green has been portrayed as a sure thing. Unfortunately, not every environmental effort produces win-win results.

Developing innovative products, bringing them to market successfully, keeping customers happy, and other elements of business success are difficult enough. Adding an environmental dimension opens up new opportunities but adds another layer of complexity to the management challenge. Gaining an edge means learning new skills, operating in new ways, and working through some hard trade-offs. In truth, the story is even more subtle. Some initiatives "fail" by traditional measures but create intangible value for a company. It's often hard to tell when hard-to-measure returns are worth pursuing.

This book attempts to bring nuance to a frequently oversimplified discussion. We dig into real-world experiences in all their complexity, highlighting pathways to success but also analyzing initiatives that didn't go as planned or absolutely flopped. We've extracted lessons, both positive and negative, from these case studies so that those now seeking Eco-Advantage don't need to start at square one.

WHY ENVIRONMENTAL INITIATIVES FAIL

Business strategies fail for many reasons, including poor planning, lack of commitment, and staffing the wrong people in key roles. But a few particular failings plague companies when they attempt to play in the environmental realm: focusing on the wrong issues, misunderstanding the marketplace, miscalculating customer reactions to green products, and failing to integrate environmental thinking fully into the work of the business. Chapter 10 reviews thirteen common stumbles that our research uncovered, but we'll touch on some of them earlier in the book to highlight the challenges of the task executives have before them.

The bottom line: Environmental initiatives take no less work than other projects. And they fail just as often. Kermit the Frog was right: It's not easy being green. But sound environmental strategy can be very rewarding.

WHO'S RIDING THE GREEN WAVE?

Defining a leader in financial performance is fairly straightforward. Pick your metric—stock performance, cash flow, or net income—and find the companies with the best or the most. Determining who the *environmental* leaders are proves far harder. Reliable data are often not available. Companies tend to measure performance in their own ways, if at all. No set of commonly accepted standards has yet emerged. Fundamentally, the environmental arena lacks the structure and rigor provided in the financial realm by the Financial Accounting Standards Board.

We began our research by trying to identify leading companies using the information available. (The details of our methodology can be found in Appendix II.) We drew on the environmental and sustainability scorecards generated by the analysts at Innovest Strategic Value Advisors, Sustainable Asset Management (which Dow Jones uses to produce its sustainability index), and others in the field of socially responsible investing. We combined these rankings with our own data, including a survey of executives. After narrowing a field of 5,000 companies to 200, we examined concrete measures of environmental impact such as emissions and energy use. This process generated a list of leaders we call "WaveRiders" (see table). While they are unavoidably incomplete, these rankings provided a starting point for our in-depth company reviews and interviews.

We explored several dozen of these top companies in detail, seeking a diversity of industry, geography, and perspective on critical environmental issues. Aware that perception plays a large part in the rankings, we added to the mix a few companies that were too small to be noticed generally but were well-known in industry circles—outfits such as the Swiss textile manufacturer Rohner Textil. We also sought out leading companies that tend not to trumpet their environmental friendliness, like furniture maker Herman Miller or cell phone giant Nokia. Finally, we made a point of speaking to some companies, like GE and Coca-Cola, that have not been considered leaders but now are either expressly seeking Eco-Advantage or have elements of their operations that are top-notch and worth studying.

We've made no attempt to study *every* WaveRider. Some industries

Top 50 WaveRiders

	United States	International
1	Johnson & Johnson	BP
2	Baxter	Shell
3	DuPont	Toyota
4	3M	Lafarge
5	Hewlett-Packard	Sony
6	Interface	Unilever
7	Nike	BASF
8	Dow	ABB
9	Procter & Gamble	Novo Nordisk
10	SC Johnson	Stora Enso
11	Kodak	Philips
12	Ford	Bayer
13	IBM	Holcim
14	Starbucks	STMicroelectronics
15	Intel	Alcan
16	Xerox	Electrolux
17	McDonald's	Suncor
18	GM	Norsk Hydro
19	Ben & Jerry's	Henkel
20	Patagonia	Siemens
21	International Paper	Swiss Re
22	Alcoa	AstraZeneca
23	Bristol-Myers Squibb	Novozymes
24	Dell	IKEA
25	United Technologies	Ricoh

are so well represented that it would have been repetitive to spend time with all the companies. Other stories are too unique to yield guidance for the typical business. Patagonia, for example, is arguably the most environmentally focused company in the world, but it's owned almost entirely by the founder, Yvon Chouinard, who prides himself on putting values ahead of profits. Indeed, he often jokes that he never wanted to be a businessperson.

We did not shy away from companies facing significant environmental challenges due to the inherent demands of their industries. Many of the WaveRiders are still big polluters—some of the biggest in the world. But they have smaller footprints than others in their industries. We believe that relative position matters. As long as the demand exists for energy, chemicals, and metals, we think it's valuable to highlight those who do these dirty jobs best. But calling them leaders doesn't mean that their work is done. In many ways it's just beginning.

The Bottom Line

Have WaveRiders been hurt by their focus on environmental matters? Much ink has been spilled trying to prove or disprove the connection between environmental and financial performance. We don't want to add to wild claims in either direction, but when we examine the stock performance of the publicly held WaveRiders versus the market overall, the trend is clear (see chart). These companies have easily outperformed the major indices in the last ten years. (And yes, Dell is on our list. But even without this high-flyer, the value of WaveRider stocks has outpaced the market.)

A word of caution: correlation is not causation. The relative stock market success of our WaveRiders might well be a function of high-quality management generally rather than any specific green focus. Indeed, a number of studies have demonstrated that environmental performance is a powerful indicator of overall management quality.

A Sustainable Path?

No company we know of is on a truly long-term sustainable course. Three additional caveats are thus required:

- All of the WaveRiders are polluting and depleting the world's natural resource base to some extent.
- Many of the companies we highlight as leaders come from industries with serious environmental impacts, but the WaveRiders are "best in class" or have practices from which others can learn.
- Not every environmental investment the WaveRiders have made has paid off. In fact, all of these companies have failed at times. But, overall, their environmental focus has helped them competitively.

Stock Performance of WaveRiders

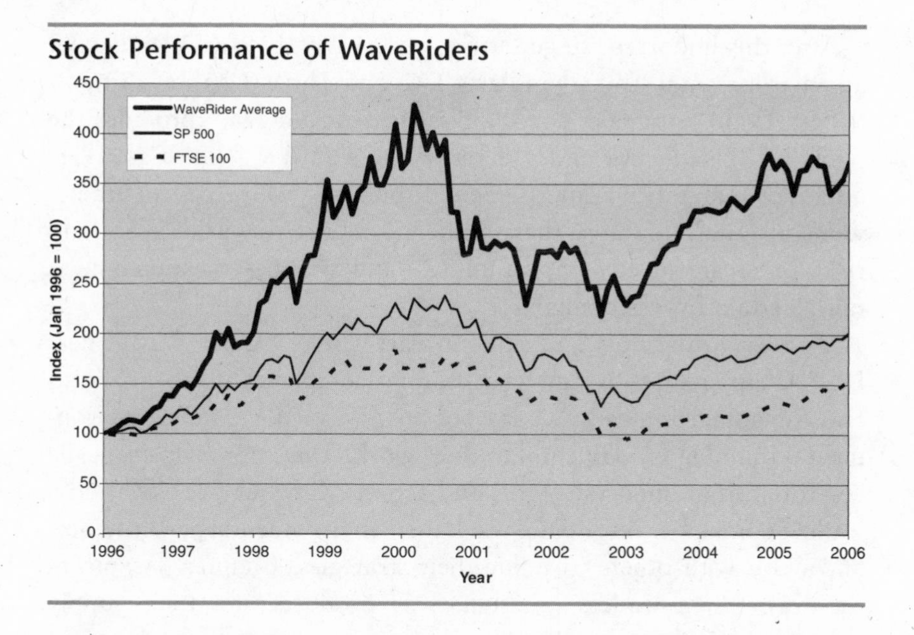

These companies aren't perfect. Some are strong on one aspect of environmental performance but weak on others. But they are all making strides. They are demonstrating a new way of doing business. In these companies, we see evidence of how to obtain Eco-Advantage—and the beginnings of a shift toward a world where environmental protection and business success go hand in hand.

THE PATH AHEAD

The chapters in Part One, "Preparing for a New World," describe the business playing field that increasingly makes environmental thinking both profitable and necessary. Chapter 2 spells out the major environmental challenges that every company faces. It reviews the state of the debate on ten major environmental problems facing the business world and humanity. In Chapter 3, we turn to the stakeholders who care about these issues and can affect a company's fortunes deeply. We lay out a framework for thinking about five groups of players, including watchdogs (such as NGOs), agenda setters (think tanks), business partners (customers), communities, and investors (banks).

After this important stage-setting, we lay out the basic elements of an environmental strategy in Parts Two and Three (Chapters 4 to 9).

Part Two, "Strategies for Building Eco-Advantage," provides the playbook. Chapters 4 and 5 describe the eight key "plays" for creating advantage from environmental thinking. We provide a framework for analyzing how these strategies reduce downside costs and risks or create upside opportunities—and whether the outcome is fairly certain or less bankable.

We turn to the nuts and bolts in Part Three, "What WaveRiders Do." Chapter 6 spells out what leading companies do to foster an Eco-Advantage mindset. We lay out how WaveRiders make environmental thinking fundamental to their work. This approach sits at the core of a set of important tools and actions. Chapter 7 looks at how leading companies mine data, track their environmental performance, and work with others to refine their strategies. Chapter 8 explores the ways these leaders redesign their products and their supply chains. And Chapter 9 describes ways to build an Eco-Advantage culture that engages both top managers and line employees.

In Part Four, "Putting it All Together," we provide an action agenda for building Eco-Advantage. Chapter 10 highlights what can go wrong on the path toward environmentally-driven competitive advantage. We review common pitfalls and demonstrate why so many environmental initiatives fail. In Chapter 11, we suggest a plan of attack with short-, medium-, and long-term action items drawn from the ideas and tools throughout the book. Finally, in Chapter 12, we review all of the elements required to execute an Eco-Advantage strategy.

GREAT TO GOOD

Sometimes people need a crisis to focus the mind. Wal-Mart CEO Lee Scott was moved deeply by the devastation wrought by Hurricane Katrina in 2005. He was also profoundly proud of his company for how it helped storm victims. But the company's finest hour got him asking:

> What would it take for Wal-Mart to be that company, at our best, all the time? What if we used our size and resources to make this country and

this earth an even better place for all of us: customers, Associates, our children, and generations unborn? What would that mean? Could we do it?

Scott went on in his speech to make the business case for attention to environmental issues and concluded on a note that many corporate CEOs might find mushy. "For us," he said, "there is virtually no distinction between being a responsible citizen and a successful business. . . . they are one and the same for Wal-Mart today."

What is Lee Scott really driving at? We think it's a new definition of greatness. As business guru Jim Collins so clearly demonstrated in his pathbreaking book *Good to Great*, companies need a deep vision, culture, and commitment, as well as a set of critical approaches, to make the move to lasting greatness. What Scott and other CEOs are saying is that great, in the traditional sense, isn't really good enough anymore. GE's Jeff Immelt put it most succinctly during his ecomagination kick-off speech: "To be a great company, you have to be a good company."

Companies that are both great *and* good inspire. Customers feel strongly about those brands, and employees work harder (and have more fun doing it). Seeking Eco-Advantage is a challenging road filled with significant potholes. There's no easy route from green to gold. But the WaveRiders are showing the way.

Chapter 2 Natural Drivers of the Green Wave

In the mid 1990s, executives at consumer products giant Unilever saw a big threat to one of their product lines coming over the horizon. The supply for the frozen fish sticks business was at risk because the world's oceans were running out of fish.

Confronted with such a stark example of nature's limits, Unilever decided to take action. With Dan Esty's help, and in partnership with World Wildlife Fund, the company set up the Marine Stewardship Council, an independent body to promote sustainable fisheries around the world. The Council certifies fisheries where the total catch is limited so that fish populations do not diminish over time. To create specific incentives for fishermen to seek certification, Unilever committed to buying 100 percent of its fish from sustainable sources by 2005.

Unilever executives see this commitment—and the substantial costs involved—as a business issue, plain and simple. Said Co-CEO Antony Burgmans, "As one of the world's largest purchasers of fish, it is in Unilever's commercial interest to protect the aquatic environment from fishing methods that will ultimately destroy stocks." One supply chain manager put it succinctly: "We are not environmentalists. We are not scientists. But if we don't do anything, we will be out of business."

Such enlightened self-interest seems obvious. But getting a clear fix on pollution impacts and natural resource constraints is harder than it sounds. The voices out there on environmental issues can be loud—and sometimes shrill and unmeasured. Certain segments of the environmental community have been preaching doom and gloom for so long that the public has tuned out. In some sense, the public has less reason to listen. In the four decades since the United States woke up to the environmental challenge, smokestack industries have reduced their air and water pollution dramatically. Globally, though, the trend lines are broadly negative.

In 2005, the United Nations released its extensive Millennium Ecosystem Assessment, a comprehensive study of twenty-four natural support systems. Most are in decline. From reduced freshwater availability to soil degradation to the risk of climate change, the problems are pervasive and many require urgent attention.

Guess which national magazine wrote in 2006: "Around the world, humanity has reduced nature's capacity to dampen extremes to an astonishing degree. . . . half the world's fresh water now co-opted for human use, half the world's wetlands drained or ruined. . . . the list goes on and on." *Mother Jones* perhaps? No, it was *Fortune*, in an article about the growing dangers of climate change and our reduced capacity to handle it.

The fact that the remaining problems are not as "in your face" as the earlier issues of air and water pollution adds to the challenge. Today's threats are often difficult to see and arise from diffuse sources, such as small businesses, households, and even individual behavior. For example, although no one car generates enough pollution to cause harm, Americans' 300 million vehicles as a whole produce serious smog and substantial levels of carbon dioxide.

Issues like global warming and loss of species can seem physically remote, long-term, and not very pressing. Compared to the immediacy of the air we breathe or the water we drink, it's hard to get worked up about consequences that are far off. But long-term issues have a nasty way of sneaking up on us. Left unattended, these types of problems can prove intractable when they do arrive.

SOME CONTEXT: PROTECTING OUR ASSET BASE

In our world of modern conveniences, it's easy to forget one immutable truth: We live within nature's boundaries. The natural world is not just that pretty thing we admire on vacation. Natural resources are the assets on the planetary balance sheet. Some of this natural capital is renewable, such as forests. Other assets such as oil are being drawn down daily. Imagine a company that systematically depleted its own assets without any plan to replenish them. Wouldn't it come under tremendous fire from angry shareholders, perhaps even finding itself struggling to survive?

As Mats Lederhausen, a veteran executive at McDonald's, told us, "In a prosperous society, you really have only two assets: people—their creativity and skills—and the ecosystem around them. Both need to be carefully tended."

ENVIRONMENTAL ISSUES TO WATCH

As the business world wakes up to the fact that many natural resources are finite, a second reality is emerging in parallel: Limits can create opportunities. Companies that manage nature's bounty and boundaries best will minimize vulnerabilities and move ahead of their competitors. That much seems obvious, but turning environmental thinking into Eco-Advantage requires mastery of an astonishing range of issues. Which ones are real and pressing? Which can a company safely ignore without placing itself at a serious competitive disadvantage?

The environmental concerns that are most urgent in any particular company will vary a great deal. What's more, environmental issues evolve over time. Scientific understanding becomes more refined. Certain products, drawing on one set of natural resources, are overtaken by substitutes. Customer preferences and tastes shift. A successful manager needs to recognize the dynamic nature of the environmental management challenge.

In this chapter we offer up a quick sketch of the "Top 10" envi-

TOP 10 ENVIRONMENTAL ISSUES

1. Climate Change
2. Energy
3. Water
4. Biodiversity and Land Use
5. Chemicals, Toxics, and Heavy Metals
6. Air Pollution
7. Waste Management
8. Ozone Layer Depletion
9. Oceans and Fisheries
10. Deforestation

ronmental issues facing humanity (see box). The most important issues for any particular company will depend on its specific circumstances—industry, location, and business model. The precise order from a planetary scale is clearly up for debate, and scientists would argue over the magnitude and urgency of these challenges. Nevertheless, as a starting point for business analysis, we've ranked them in rough order of importance.

For each issue, we provide a "state of play" assessment that spells out our best scientific understanding, as well as an analysis of business consequences that highlights how the issue might affect specific companies and industry groups. By necessity, this chapter is more executive summary than exhaustive study.

CLIMATE CHANGE

Overview

No issue looms larger in terms of potential strategic impact on business than the buildup of greenhouse gases in the atmosphere. In the media, all this fits loosely under the heading "global warming," but the problems go far beyond rising temperatures. What we're facing is more accurately described as "climate change." This catch-all includes rising sea levels, changes in rainfall patterns, more severe droughts and floods, harsher hurricanes and other windstorms, and new pathways for disease. (Malaria, for example, spreads to places with warmer climates.) Without being overly dramatic, it's fair to say that climate change could threaten the habitability of the planet.

Some skeptics suggest that the science underlying this problem is not very solid. So let's review what is known and what is not by focusing on a few questions:

1. Is global warming real?
2. What and who are causing it?
3. What could climate change do to the planet and to us?
4. Who's going to be hit hardest?

The short answers are (1) yes, (2) humans, (3) destabilize our basic ecosystems, and (4) mainly the poorest, lowest-lying, and hottest countries, but everyone is vulnerable.

IS IT REAL?

Let's start with what we know. First, greenhouse gases in the atmosphere—including carbon dioxide, methane, and several other trace gases—trap heat that would otherwise bounce off the planet back into space. In fact, without the greenhouse effect, the Earth would be too cold to support life.

But humans have added substantially to this heat-trapping capacity over the past several centuries. Since pre-industrial times, the level of carbon dioxide in the atmosphere has increased from about 280 parts per million (ppm) to 380 ppm today (see figure). This may not seem like much of a difference, but it is. What we're seeing now is unprecedented. Ice core samples from the Arctic show that the greenhouse gas levels over the last 650,000 years had never risen above 300 ppm—until the modern era. Now we're facing the real prospect of a further increase to 500 or even 600 ppm over the next 50 to 100 years.

The scientific consensus that this change is caused by human activity is strong. Are there uncertainties in climate change science? Of course. The speed, magnitude, and regional distribution of effects are all matters of dispute. But out of over 900 peer-reviewed articles in scientific journals on climate change, none disagreed with the "consensus position" that climate change is real and deserving of policy attention. As *Science* magazine said, "Politicians, economists, journalists, and

Carbon Dioxide in the Atmosphere

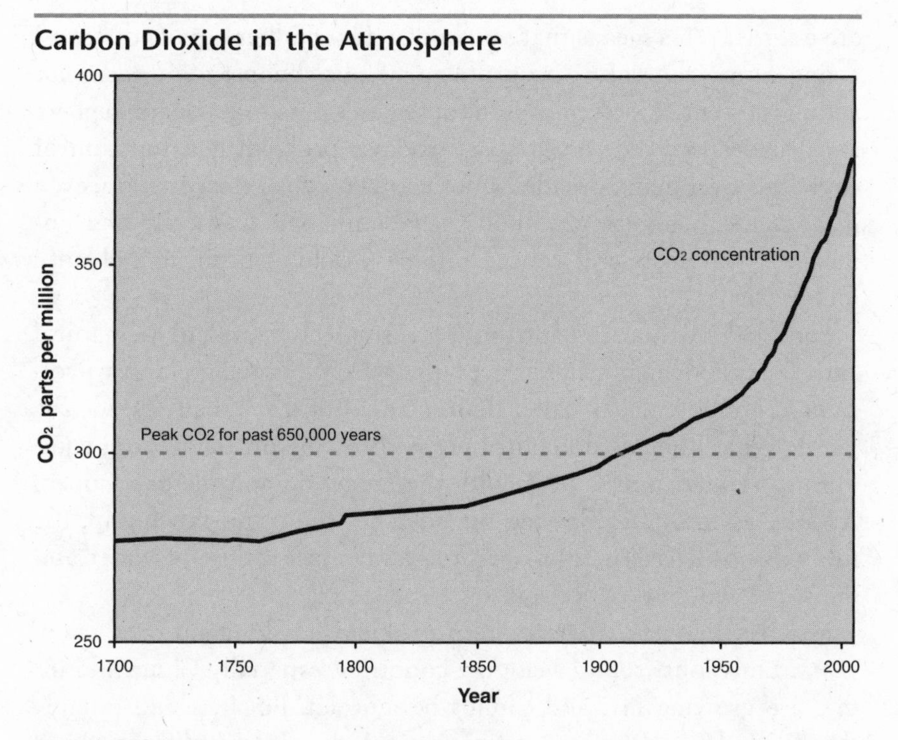

Source: Carbon Dioxide Research Group, Scripps Institution of Oceanography, University of California.

others may have the impression of confusion, disagreement, or discord among climate scientists, but that impression is incorrect."

WHO AND WHAT ARE CAUSING IT?

We know where greenhouse gases come from. Carbon dioxide (CO_2) accounts for over 70 percent of the problem. CO_2 emissions come mainly from the burning of fossil fuels. These emissions emerge from three sectors in roughly equal portions: transportation, residential and commercial, and manufacturing. The next most important source of greenhouse gas emissions is methane, particularly from natural gas leaks and off-gassing from rice paddies and flatulent cows (believe it or not). The remainder is a mixture of trace gases, including oxides of nitrogen.

Two other points about greenhouse gases bear mentioning. First, these gases blanket the earth, so it really doesn't matter where they are emitted. This means that no country (or small group of countries) acting alone can solve the problem. Global cooperation and joint action is essential. Second, greenhouse gases persist in the atmosphere for decades or even centuries, so today's problem is a function of emissions over many decades. On the other side of the equation, even if we were to reduce greenhouse gas emissions drastically and immediately, atmospheric concentrations wouldn't begin to fall until mid-century.

Some say the human contribution is small compared to the natural carbon cycle, which includes everything from decaying plants to volcanic eruptions. Technically, that is true. But it's actually beside the point: What humans do contribute is throwing off the natural equilibrium. Think of a bathtub with the faucet on and the drain open. As long as the water flowing out matches the water flowing in, the tub will not overflow. But open the faucet just a tiny bit wider and the water will eventually spill over the sides.

We also know that the current problem has built up over many years. Emissions from developing countries, especially China and India, are growing fast and cannot be ignored. But it is indisputable that the industrialized countries, particularly the United States, have caused the bulk of the existing problem.

WHAT WILL HAPPEN TO THE PLANET AND TO US?

To be honest, no one knows exactly what climate change will do. But we've got some pretty good theories and forecasts:

Hotter temperatures. Nine of the ten hottest years on record happened after 1995, with 2005 the hottest ever. While the average temperature increase may seem small, much of the strain comes from higher and more frequent temperature spikes. The 2003 heat wave in Europe killed 26,000 people, and half the summers in Europe could be that hot by 2040.

Rising sea levels. Ice is melting all over the world. The U.S. Rockies have seen a 16 percent decline in snowpack, and Africa's Mt. Kilimanjaro has lost 80 percent of its ice in recent years. The potential result of some of this melting, specifically in Greenland and Antarctica, is a significant rise in sea level. Large portions of many low-lying countries could be permanently flooded, and coastlines could recede well into coastal communities worldwide.

Increased intensity of windstorms. A year before Hurricane Katrina slammed the Gulf Coast, the *New York Times* reported a widespread belief among scientists that climate change is noticeably increasing the intensity of hurricanes. Whether any single mega-storm, such as Katrina, is a signal of climate change remains unclear. As ocean temperatures rise, however, the prospect of more severe hurricanes and typhoons cannot be doubted.

Disrupted ecosystems. Fundamental changes in what grows where and which species die or thrive loom on the horizon as temperatures rise, rainfall patterns shift, and hydrological (water) flows change. Maple trees may no longer survive in Vermont. Warmer temperatures in British Columbia and Alaska have already allowed the pine bark beetle to thrive where it once could not, destroying millions of acres of forests. Some forecasts suggest that parts of the U.S. Midwest may become as drought plagued as it was in the Great Depression, dramatically reducing productivity across America's breadbasket.

Displaced peoples and environmental refugees. The tragic scenes from post-Katrina New Orleans showed us the chaos and suffering that often results when thousands of people are suddenly forced to move. In the coming decades, we may see a forced exodus from both small Arctic villages and large swaths of countries like Bangladesh, where tens of millions of people live just above sea level. Where will all these people go?

WHO WILL BE HIT HARDEST?

Countries without the capacity to handle drastic change will likely not fare well. Low-lying countries, especially those close to the Equator, face especially significant hurdles. The most exposed nations are often among the world's poorest, making the potential consequences all the more dire. Parts of the United States are also at risk. Low-lying areas in the hurricane belt, including much of the southeastern United States, will be exposed to increased wind damage and greater storm surges.

Business Consequences

Climate change is a highly contentious issue because the costs both of addressing the problem and of *not* addressing it could be very high. Greenhouse gases are linked to the burning of fossil fuels, so successfully controlling emissions will require action on the part of not only every business on the planet but every individual.

Rising temperatures and more unpredictable weather could affect a broad range of livelihoods and industries. Changed rainfall patterns could be devastating to farmers. Ski resorts could melt away. And more severe storms could wreak havoc on transportation systems and airlines.

Some businesses are already feeling the effects profoundly, such as the insurance industry. The economic cost of extreme natural disasters has increased tenfold since the 1950s. In the face of further risk and uncertainty, reinsurance companies, like Swiss Re and Munich Re, are actively pushing for action on climate change.

Beyond the direct weather and temperature effects, every company will face the second-order effects of climate change, particularly as regulatory policies to control greenhouse gas emissions kick in. Carbon charges and higher fuel prices will require redesigning distribution systems, supplier relationships, and many other aspects of business operations. For airlines, logistics companies, and any business with heavy transportation needs, vehicle fuel efficiency will become critical. Those industries that rely on petroleum feed stocks, such as chemical and plastic manufacturers, will need to rethink their materials use strategies.

Regulations limiting greenhouse gas emissions are already a reality in Europe and Japan. Most observers of the policy process expect that carbon constraints will be imposed in the United States in the next few years. Indeed, at a recent summit of power company leaders, a GE executive asked how many of his peers thought that a mandatory cap on greenhouse gas emissions was coming, no matter which party wins in the post-Bush era. Nearly all raised their hands.

Changes on this scale create large opportunities. Consumers, communities, and businesses will need new technologies, products, and services as they adapt to climate change and a carbon-constrained world. Plenty of smart companies have the product lines and ingenuity to make money filling this need. As households and businesses move to reduce their energy consumption, for example, a company like Honeywell can offer sophisticated thermostats as well as efficient heating and air conditioning equipment. And those who provide technologies that mitigate the problem at its source—with greenhouse gas–free energy supplies, for example—will clearly profit. Likewise, any company that demonstrates cost-effective ways to capture and hold carbon dioxide instead of letting it reach the atmosphere, called "carbon sequestration," will win big markets very quickly.

THE CLIMATE CHANGE IMPERATIVE

Climate change is shaping up to be the biggest environmental strategy issue the business world has ever faced. The potential effects are both broad and substantial. The need to rethink strategy with an eye on climate change impacts and regulatory constraints is fast becoming a corporate imperative.

ENERGY

Overview

In 2005, ChevronTexaco began running full-page ads in major magazines, declaring, "the era of easy oil is over." A high-ranking official at another major oil company told us that world-wide oil production would peak in the next few years. Geologically, oil gets more difficult and more expensive to extract every day. Whether or not climate change policies drive us away from fossil fuels in the short run, the bottom line is clear: the energy future will not be the same as the energy past.

Energy is not exactly an environmental problem like the others on our list. But every society needs energy, and energy production, no matter what the method, can damage the environment. Fossil fuel burning, of course, causes pollution and contributes to the build up of greenhouse gases in the atmosphere. Even "clean" energy sources like hydro power have their own environmental consequences like changed groundwater flows and obstructed fish migration.

Some forecast an end to what energy guru Dan Yergin has called the era of "Hydrocarbon Man." Others dispute this claim. While acknowledging that a carbon-constrained world is upon us, they envision new technologies that will allow the capture of carbon dioxide and dangerous pollutants like mercury, thereby enabling fossil fuel burning for years to come.

Clearly the exact path of our energy future is yet to be charted, but whatever course it takes will affect every industry. To cite only one example: At present more than two-thirds of U.S. electricity is generated from fossil fuel combustion—51 percent from coal, 17 percent from natural gas, and 3 percent from oil. With energy demand continuing to rise, particularly in the fast-growth areas of the developing world such as India and China, the price of such fuels will likely remain high for the foreseeable future.

That's the bad news. The good news is that oil prices above $50 per barrel transform the energy marketplace. Renewable energy sources such as wind, solar, geothermal, bio-based fuels, and tidal power become increasingly price competitive. In some regions, wind power has already begun to capture a noticeable market share. Although the levels of current production are low, the growth rates of the renewable energy industries are impressive. Global wind power capacity has been growing at over 30 percent per year, and solar power at over 60 percent. Simultaneously, support for nuclear power has been rising, even in some environmental camps, because it produces no local air pollution and no greenhouse gases. Nuclear plants remain highly controversial for security and waste disposal reasons. But the economics of energy can create both new opportunities and odd bedfellows.

Some energy commentators, as well as public figures like Governor Arnold Schwarzenegger, have heralded the arrival of a hydrogen economy as the solution to both our energy and climate change problems. But hydrogen is really a way of storing and distributing energy, not an energy source. As a fuel for vehicles, hydrogen would ensure zero *local* emissions, a boon to cities. Electricity would still be needed, however, to generate hydrogen from water. Thus, hydrogen is only zero-emissions if the electricity to produce it comes from a renewable non-emitting source such as wind or solar power. Hydrogen may be an important part of the energy mix down the road, but we are many years from having an affordable mass-market system for converting, transporting, and delivering it.

Business Consequences

Inevitably, the changing energy picture creates new competitive pressures. For big energy users—those in heavy manufacturing or transportation, for example—resource and energy productivity may become a major point of strategic advantage. With higher prices a near certainty, incentives for energy conservation will rise, and energy efficiency investments will pay substantial dividends. Companies selling goods and services that promise to improve energy efficiency will claim market share. For example, Tide has released a new cold-water detergent with advertisements touting the substantial energy savings from not having to use hot water.

The demand for renewable energy is also rising. Many large U.S. states, including California and the entire Northeast, now mandate that utilities provide a certain percentage of their supply through renewable energy. Some companies are following suit without legal pressure. Microprocessor manufacturer AMD has committed to purchase renewable energy to power all its Austin, Texas, facilities for ten years. Companies as diverse as Starbucks, FedEx Kinko's, and Johnson & Johnson buy 5 to 20 percent of their energy from renewable sources. Most strikingly, Wal-Mart has committed to having 100 percent of its energy come from renewable sources.

Alternative energy companies are hot investments as venture capitalists large and small envision an energy future that looks quite different from the past. Major Silicon Valley players, including the celebrated investors Kleiner Perkins, have started "clean tech" funds to invest mainly in renewables. The share of venture capital going into these funds has been rising rapidly (see figure). By all accounts, these markets for environmental investments will be extremely large: a United Nations report indicates that money flowing into clean tech will be in the trillions over the next decade. From solar cells to wind turbines to hydrogen vehicles, entrepreneurs are working to make their solution the energy source of the future. Like all new ventures, not all of these businesses will succeed. But some will profit mightily.

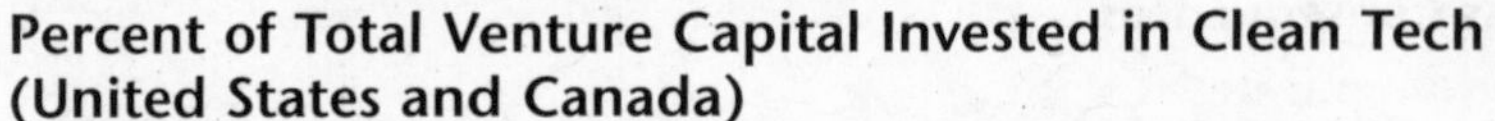

Percent of Total Venture Capital Invested in Clean Tech (United States and Canada)

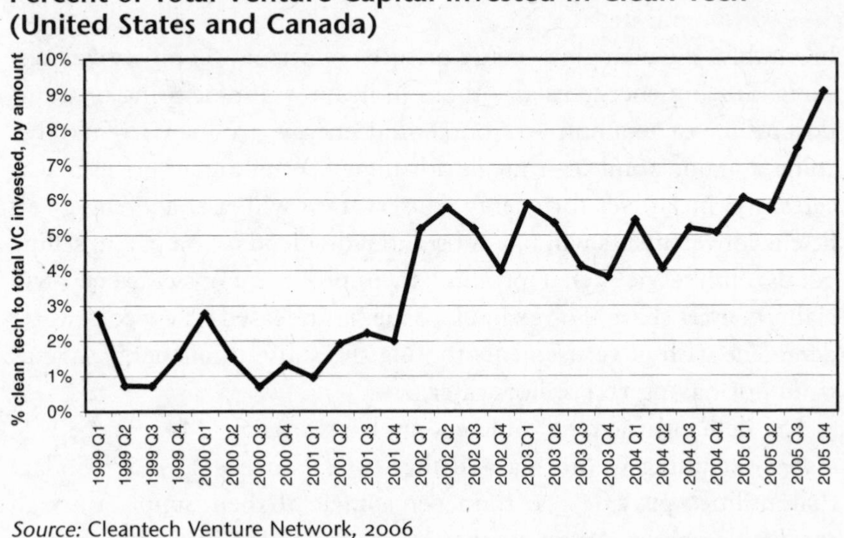

Source: Cleantech Venture Network, 2006

WATER

Overview

Water is the essence of life. It's also a critical input to agriculture and many industrial processes. Companies around the world now face real limits on access to water. A rising population and growing economies are putting substantial stress on resources in drier regions. Even where water is relatively plentiful, water pollution is increasingly a concern. For business, these multiple, complementary factors create both water *quality* and water *quantity* challenges.

QUALITY

Everywhere you look governments and communities are growing more concerned about protecting the quality of drinking water supplies. In the United States and other developed nations, quality has generally improved in the past several decades. Long gone are the days where waterways were so contaminated that they might catch on fire, as the Cuyahoga River in Cleveland did in 1969.

But our waters are still threatened by industrial effluent, agricultural runoff, and contamination from sources as diverse as mining operations, construction sites, and our lawns. Governments across the world continue to mandate ever-lower flows of pollutants into rivers and streams. Companies must plan for increasingly stringent water pollution standards.

In the developing world, a dismal 90 to 95 percent of all sewage and 70 percent of industrial waste goes untreated directly into rivers, lakes, or the ocean. Protecting precious water supplies has emerged as a priority in every country.

QUANTITY

The Millennium Ecosystem Assessment estimates that as many as two billion people live with water scarcity. It's a central issue in drier regions the world over. The problem boils down to simple supply and demand. The Earth is a closed system, so freshwater supply is basically fixed. But rising populations and an increase in irrigated crops continue to drive up water demand. Indeed, agriculture consumes about 70 percent of the water we use. In many regions, we use more water than the natural rainwater cycle provides. As a result, nature's underground water supplies, called aquifers, are being drawn down. The giant Ogallala Aquifer, for instance, which lies under parts of eight U.S. states and holds more water than Lake Huron, has been dramatically depleted. Once again, we are drawing down our natural capital.

The problems of water quantity are already acute in China. Hundreds of cities face potentially severe shortages within the next few years. The water table under some of China's main wheat-growing regions is dropping precipitously. In just six years, grain production has fallen by 70 million tons (equal to Canada's total crop), driving up food prices for people who can't afford it.

These water quantity concerns may actually be understated. There's a growing recognition that immediate human needs are not the only ones that matter. Nature also needs water to support plants and animals, which in turn support us. The conflict between human demands and ecological needs can get ugly at times. In 2002, more than 34,000 Chinook and Coho salmon died in Oregon's Klamath River as the water flow, drained by expanded irrigation, fell below the level needed to support fish life.

10-SECOND PRIMER: WETLANDS

A relatively new issue on the corporate radar is the need to protect wetlands. In the bad old days (pre–1989), the Army Corps of Engineers would drain what they then called swamps to prevent disease or control flooding. We now see wetlands as a vital ecosystem that provides life support and habitat for plants and animals. For humans, they filter pollution and control storm surges. (The loss of wetlands in the New Orleans area is one reason the 2005 flooding was so catastrophic.) Companies must now manage their land resources and any development plans with careful attention to wetlands preservation.

Business Consequences

Companies must expect increased scrutiny of their water use. Those deemed to be using too much water or degrading water quality will face political attack, public backlash, intensified regulation, and even legal action.

For Coca-Cola, it's impossible to disentangle the company's license to operate from how it handles water. In Kerala, India, the local government shut down Coca-Cola's bottling facility for two years out of concern about the plant's water consumption. Coke isn't alone. Many companies depend on water and are growing more active in managing the problem. As Unilever's environmental report recognizes, "Working with consumers to foster the responsible use of water is clearly in our long-term interest . . . because without clean water many of our branded products would be unusable." Put simply, people can't wash clothes using Unilever detergent if there is no water.

Water quality concerns are a world-wide issue, not just a luxury item for rich countries to worry about. Just ask the managers at Celulosa Arauco's $1.4 billion Valvidia pulp and paper mill in Chile. In the past, untreated waste entering a local wetlands might not have seemed like a big deal. But when the company's effluent was implicated in the deaths of thousands of swans, the government shut the mill down for a month, costing the company over $10 million. As the tragedy unfolded, a series of executives were demoted or ousted, including the CEO and environmental manager. Ultimately, the company had to shut the plant down for another two months to fix the problems.

Coca-Cola, Unilever, and many other companies operating in the developing world are trying to manage downside risks. On the positive side of the ledger, companies that provide solutions to water problems may gain favor with local communities. Some may even profit as they identify new market opportunities. Over the last few years, mainly through acquisition, GE has built a multibillion dollar water infrastructure business. With these investments, GE has positioned itself to "solve the world's most pressing water reuse, industrial, irrigation, municipal, and drinking water needs." GE has clearly spotted the Green Wave and wants to ride it.

BIODIVERSITY AND LAND USE

Overview

"Our personal health, and the health of economies and societies, depends on the variety of ecological goods and services that nature provides." A statement from Greenpeace? No, it's from mining giant Rio Tinto.

Biodiversity—a catch-all term for the spectrum of plant and animal life around us—preserves our food chain and the ecosystems on which all life depends. It also holds the prospects of new drugs, foods, and other products derived from newly discovered species. By its very nature, biodiversity is hard to measure, but it's increasingly considered a critical natural resource that society must manage like any other. As a ballpark figure, one respected team of researchers put the value of biodiversity and ecological services in the trillions of dollars per year.

SIGNS OF DECLINE

How serious is the problem of biodiversity loss? The U.N.'s Millennium Ecosystem Assessment affirmed one scientific hypothesis: The extinction rate is now as much as 1,000 times higher than the average rate over Earth's history. Over the last five billion years, biologists tell us, the Earth has gone through five major tumultuous periods with rapid evolution and massive species purging. We have now entered the sixth major extinction period, but this is one of our own making.

The main problem is the pattern of human development, which tends to destroy natural habitat. Toxic chemicals and pollution also play a key role, and climate change is rising on the list of causes. Recent evidence shows that the pending extinction of many frog species—the canaries in the coal mine of biodiversity—is caused by warming temperatures.

The best ecological scientists estimate that one-quarter of the planet's species are in some danger of extinction within a single human generation. We may live to see the end of the polar bear, many of the big cats, including the majestic tiger, and millions of less charismatic species. Some casual observers might mock concern over lost insects and micro-organisms, yet these species play a critical role in the complex web of life. The not-so-funny joke is that bugs and bacteria could get along fine without us, but we wouldn't survive long without them.

A related problem is the introduction of non-native plants and animals into ecosystems. These "invasive" species, often arriving as a result of trade and shipping, can threaten ecosystems, damage native wildlife, and cause of millions (even billions) of dollars of harm. For example, zebra mussels in the Great Lakes clog water intake pipes and damage boats. Congressional researchers estimated that these pesky mollusks cost the power industry alone more than $3 billion in the 1990s.

LAND USE

A key factor in the decline of biodiversity is habitat loss. With population growth fueling more land use and a rising standard of living, pressures on ecosystems are seemingly unavoidable. But our development choices have often made matters worse. Suburban sprawl, with little attention to the preservation of open space, has fragmented ecosystems, jammed highways, and diminished the quality of life for humans and animals alike. In the developing world the problem is very different, but no less acute. The development pressure comes from land conversion for crops and slash-and-burn agriculture. It's a matter of survival for the world's poor, but the price is high in terms of lost biodiversity.

With open space in decline, environmentalists are pressing governments to preserve wild places by creating new parks. A number of companies have contributed land and resources in support of these efforts. Wal-Mart, in a bid to overcome its image as the #1 engine of sprawl, launched a program called Acres for America. The company committed to preserve land equal in area to what it says is its entire operational footprint, over 130,000 acres, providing something of an offset for the land it has consumed. This action isn't a full environmental strategy by any means, but it is one attempt to answer the concern some communities have about how Wal-Mart gobbles up land.

Business Consequences

Many companies face pressure about their contribution to sprawl. Communities are recognizing that they must plan land use carefully to avoid breakdowns in transportation systems and a reduction in quality of life. Companies in the densest areas of development—northern Virginia, Florida, metropolitan Atlanta, California, and the entire Southwest—are justifiably worried about how they'll get their goods and even their employees where they need to be.

Wherever companies locate factories or stores, communities and governments are demanding more careful consideration of the damage to ecosystems. Those who use natural resources most intensively, including real estate developers, mining enterprises, and timber companies, will face the biggest scrutiny. (In Chapter 6, we'll see how Rio Tinto is proactively managing biodiversity issues to reduce enterprise risk and improve access to new lands.)

Aside from the drug industry, which harvests new compounds from plants and animals, the upside of managing for biodiversity is more speculative and the stories fewer. But the potential is big. Billions of years of evolution have produced some stunning products for human technology to copy. Spider silk, for example, is five times as strong as steel, yet stretches. A spider-silk net, if it were big enough, could stop a jet in flight without breaking.

It's no surprise, then, that some companies are quietly studying animals and plants for what we can learn about how they operate. When Interface Flooring sent its designers out to the forest to study how the natural world is designed, they noticed extraordinary similarities in nature's patterns, but no exact repetition. The company then captured that feel by creating a new line of carpet tiles with varied designs that can be placed in any order on the floor and still make a coherent whole. This natural look allows for lower production costs and much faster installation. The new brand, called Entropy, became Interface's fastest selling product ever.

From an economic point of view, biodiversity and natural habitats have real value, though the precise economic benefits are difficult to quantify. Forests, for example, provide more than wood: They purify polluted water and reduce floods. We see a growing interest in the public and private sectors in finding ways to measure and price the "ecosystem services" the natural world provides. Even without exact measurements, some governments have taken action. China banned logging near areas where flooding has become a problem. They clearly valued the flood protection more than the forestry revenues. And instead of building a $4 billion water filtration plant, New York City spent only $600 million to buy land in the Catskill Mountains to preserve the natural water purification services.

The private sector is also getting in on the act. In an effort to keep its water pure, Perrier subsidizes the reforestation of watersheds and pays farmers near its water sources to go organic.

Finally, efforts to control the effects of invasive species are gaining momentum. From the Asian long-horn beetle to zebra mussels, these stowaway species promise to add to shipping costs for anyone moving goods internationally as governments pass on the burden of programs to control the problem.

CHEMICALS, TOXICS, AND HEAVY METALS

Overview

Part of what makes air pollution—and all forms of pollution—more dangerous is the presence of toxic elements. Exposure to chemicals like dioxin, a by-product of production processes such as papermaking, and heavy metals such as lead and mercury can create severe public health risks. The fear of cancer or possible birth defects has led to stiff chemical control laws in both the United States

10-SECOND PRIMER: THE PRECAUTIONARY PRINCIPLE

The "precautionary principle," a bias toward safety in the face of uncertainty, lies at the heart of the EU approach to environmental regulation. "Better safe than sorry" seems like common sense to many. Others note that, in practice, the precautionary principle works as a bias toward the status quo. It may also deter innovation or be used by "protectionists" to shield existing producers from new competition.

and in Europe, where a strict precautionary principle is in place (see box).

The European Union's new REACH directive mandates that manufacturers, *not* regulators, must prove the safety of every new and old chemical—a complete reversal in the burden of proof. The law is not without controversy, but given the public support for heightened protection, it's likely to stand.

The legal liability surrounding toxics can turn out to be virtually unlimited. For years, scientists knew that exposure to asbestos was dangerous, but the legal system is just now catching up with the science as judges and juries are imposing sizable damages on companies that produced asbestos. More than 70 companies have been driven into bankruptcy, including longstanding businesses such as Johns Manville and W.R. Grace.

New sources of concern could easily follow the path of asbestos, becoming significant sources of legal exposure in the years ahead. Endocrine disruptors, chemicals used in everything from insecticides to detergents to plastics, may change hormone levels in animals and people, and thus throw biological processes such as reproduction, growth, and immune function out of whack. The amount of scientific research and critical attention on how these chemicals affect us is rising fast. One telling sign: The annual number of articles in medical journals studying the health effects of hormonal agents rose from 200 in 1990 to nearly 1,000 in 2002.

Another relatively new concern, at least in the United States, are chemicals in flame retardants called PBDEs. These chemicals were adopted for public safety reasons and applied to a wide range of products from electronics to clothing. We now know that PBDEs are neurotoxins that potentially affect learning and attention and can be passed along through breastfeeding. Breast milk studies show concentrations 20 to 100 times higher in U.S. women than in countries where they've been banned. California has decided to ban PBDEs by 2008.

HEAVY METALS

The science on lead could not be more clear. The metal can cause brain damage and severe developmental problems in children. Its elimination from gasoline, paint, water systems, and other pathways of human exposure is one of the great public health success stories. Governments are now applying the same regulatory rigor to mercury, cadmium, and other heavy metals.

One particular area of concern is the airborne mercury released from burning coal. The vaporized mercury settles in waterways and from there enters our fish, our food, and us. The United States houses over 600 coal-fired power plants and thousands of smaller-scale boilers. Activists are pressing the U.S. Environmental Protection Agency (EPA) for action, and stronger standards on mercury releases are a near certainty in the coming years.

Mercury is one of the main reasons for recent EPA fish advisories, warning the most sensitive populations such as pregnant women and children against even moderate consumption of certain types of fish. And the concern about mercury is justified. A study from Mt. Sinai Hospital and Albert Einstein Medical School in New York suggests that prenatal mercury exposure increases the risk of reduced brain function and developmental problems in 630,000 children each year, at a cost to society of $8.7 billion. Roughly 10 to 20 percent of women of child-bearing age have mercury levels in their blood in excess of EPA health standards.

Business Consequences

Companies today must pay careful attention to what they produce and how they produce it. Products containing toxics or heavy metals face special regulatory hurdles. Failure to track these rules carefully can lead to trouble, as Sony found out during its PlayStation Christmas debacle. The cost of managing chemicals in production pro-

EVEN WAVERIDERS CAN TAKE A TUMBLE

Chemicals and toxics are complicated issues to manage correctly, with ugly consequences for companies that get caught on the wrong side. DuPont has recently fought charges that it ignored the dangers of a chemical called PFOA, which is used to make Teflon. In 2004, the company agreed to a $100 million–plus settlement in a class-action law suit involving PFOA-contaminated drinking water in Ohio and West Virginia. The company also faced a U.S. EPA accusation that it had withheld information about the safety of the chemical. DuPont paid $16 million in fines and penalties to settle the charges. Notwithstanding the fact that PFOA is unregulated, DuPont concluded that it was not worth continuing the legal battle given the damage being done to its reputation.

cesses, including the disposal of toxic waste, can be very high. Toxics can cost a business in many ways; just try to unload a contaminated property, for example.

Tracking chemical use carefully, with an eye toward the redesign of products and processes to eliminate unnecessary toxins, is often a good investment. Computer and printer giant HP, for example, is removing flame retardants from its products ahead of the likely regulations.

Moreover, concern is growing about whether chemical exposures are contributing to an increase of certain diseases and conditions, particularly in children. Is the rise in rates of autism, childhood cancer, and certain allergies related to toxics? It may be that doctors are better at diagnosing these conditions now than in the past. But some parents wonder if chemicals in our food are playing a role. And when parents become emotional, pressure always mounts on companies and governments to manage the chemicals more closely.

The potential legal liability for companies producing and using toxic substances or heavy metals is very high. And the potential for "court of public opinion" decisions that condemn companies that now know better could be even higher. That's why cosmetic giants like Revlon and L'Oréal agreed to remove the chemicals called phthalates from their cosmetics.

On the upside, we see an opportunity to feed the growing demand for healthy alternatives in everything from food to cosmetics. Whole

Foods, a leader in providing a health-seeking public with fresh organic foods, is one of the fastest growing supermarkets in the country. What was once a niche play is now turning into a large market opening.

AIR POLLUTION

Overview

Not long ago, commuters couldn't see six traffic lights ahead in Los Angeles. Today air quality is not perfect, but it's much better. Significant air quality controls on factories, cars, and other emissions sources have radically reduced air pollution levels over the past thirty years in the United States, Japan, and Europe. We all can breathe a little easier.

The gains have also helped ecosystems, like the forests and lakes in the northeastern United States that suffered from acid rain in the 1980s. One of the big success stories in recent environmental policy is the vast reduction of sulfur dioxide emissions in the United States. The 1990 Clean Air Act amendments, signed by the first President Bush, created a free market for companies to trade sulfur dioxide emissions allowances. The program proved to be a bigger success than anyone predicted. The concentration of sulfur dioxide in the atmosphere is less than half of what it was—all for a fraction of the cost originally projected.

While our air is clean*er*, it is still not clean in many places. Serious health risks remain for people who suffer from respiratory illnesses like asthma. The European Commission estimated that air pollution causes over 300,000 premature deaths in Europe and costs $100 billion in lost work time.

In the developing world, outdoor—and especially indoor—air pollution remains a very serious problem. Indoor air pollution, caused mainly by open fires for heating and cooking, creates particularly severe health risks. In the developed world, indoor air pollution created by chemical relases from furniture, carpet, and paints is a factor in so-called sick building syndrome.

Business Consequences

Controls on air emissions will remain strict across the board. Standards and enforcement are likely to tighten modestly in the developed world while rising substantially in developing countries. Companies with potentially significant air pollution impacts must plan to face this challenge. It is no surprise that those in dirty businesses, like cement, are at the front of the parade. Lafarge, for example, has led its industry into a major sustainability initiative with the Geneva-based World Business Council for Sustainable Development.

As air quality factors are better understood, the air quality issue affects a more diverse group of companies. Regulations that used to apply only to the big guys now also apply to smaller businesses. In addition, companies that make indoor products treated with chemicals are hearing some noise from customers. Questions from the marketplace are a big reason that WaveRider Herman Miller spends a quarter million dollars annually testing its office furniture for off-gassing.

WASTE MANAGEMENT

Overview

In the 1970s, the "don't pollute" messages—and the crying Native American on TV—alerted the public to the problem of litter and waste. Suddenly, everyone began to worry about garbage. A generation grew up understanding that we had too much waste and needed to recycle, and the results have been dramatic. Today, the United States recycles about 20 percent of its glass, 40 percent of paper, 50 percent of aluminum, and 60 percent of steel. Other countries have done as well or better. Sweden recycles 90 percent of its glass and aluminum, and Japan 86 percent of its steel.

But the disposal of wastes from factories, offices, and residences is still a challenge. By volume, the largest part of the problem is solid waste—the everyday materials discarded by homes and offices—that we either incinerate or dump in landfills. Although smaller in volume, toxic waste presents a bigger management challenge. In most developed countries, a manifest system requires the tracking of waste in great detail, often from "cradle to grave." Indeed, the failure to keep track of toxic waste can be a serious crime. In the United States, more people go to prison for mishandling hazardous waste than for any other environmental issue.

An emerging problem is what to do with all our outdated electronics equipment. This "e-waste" is becoming a burden for countries and companies alike. Every older computer has about four pounds of toxic materials including a who's who of the worst offenders—flame retardants, lead, cadmium, and mercury. Given that over 300 million computers are awaiting disposal in the United States alone, the toxic waste math is not pretty.

Business Consequences

What's behind the now-famous business mantra, "Reduce, Reuse, Recycle"? We see three main issues. First, it's often the law. No business can afford to be careless with hazardous waste. The penalties—from possible jail time for managers to multimillion dollar liability settlements—are just too great.

Second, waste reduction can cut costs. Companies save money on landfill tipping fees when they produce less garbage. Hazardous waste disposal costs can be ten times greater. Many of our Wave-Riders have drastically reduced waste by redesigning products and processes and by increasing recycling efforts. Furniture maker Herman Miller has even set a goal of zero waste to landfill, and fully

10-SECOND PRIMER: SUPERFUND

In older industrial areas, the cleanup of abandoned hazardous waste sites remains a big issue. The EPA estimates that the 1,200 Superfund sites across the country will require about $200 billion to clean up over the next thirty years. As anyone who's seen the movie *A Civil Action* knows, under the liability provisions of the Superfund law, anyone found responsible for the waste at a site can be held liable for the full cost of cleanup, even if the toxics were disposed of legally. Escaping liability for cleanup costs is difficult, and the penalty for mismanaging Superfund issues can be very high.

expects to meet it. This is a monumental feat, and not every company can pull it off. But many companies can save a lot of money just by producing less waste, and not only because they're going to the dump less often. Too many companies are throwing out valuable stuff. If they capture it instead, they will save both money and materials.

Third, new "extended producer responsibility" laws, particularly in Europe, are forcing industries like electronics to design out some elements or take their products back and handle disposal themselves. As a result and in anticipation of such laws spreading globally, smart companies are developing "take-back" programs. Nokia, for example, saw the emerging legislation and developed an extensive program in Europe, ahead of regulations.

Apart from reducing cost and risk, some businesses are seeing upside potential—a chance to profit from recycling and reuse. The *Fortune 100* list of the fastest growing companies includes firms like Schnitzer Steel Industries and Steel Dynamics that recycle scrap metal into steel—very "old economy" but very high growth potential. A number of bigger companies are trying to head off the niche recyclers and keep the profits from the end of the value chain for themselves. Xerox reclaims toner cartridges in part to keep others from refilling and reselling them.

In some industries, recycling helps companies reduce their contribution to nearly all of our Top 10 environmental problems. Take aluminum production, one of the dirtiest and most energy-intensive processes in modern society. More aluminum recycling means less need for virgin ore, less smelting, and fewer new mines. These reductions mean lower greenhouse gas emissions, less toxic runoff (from mining), and reduced land use and biodiversity problems. In short, recycling in the aluminum industry greatly reduces the burden on air, land, and water—a win-win situation for everyone.

OZONE LAYER

Overview

In what became worldwide news in the 1980s, a hole in the planet's protective ozone layer opened up over Antarctica. The culprit was a set of chemicals called chlorofluorocarbons (CFCs) breaking down the ozone in the stratosphere. Like climate change, the depletion of the ozone layer is an inescapably global issue. Emissions of CFCs from anywhere spread everywhere. No country can address the problem on its own.

With a thinned ozone layer, the world becomes a more dangerous place, with reduced agricultural productivity, higher risk of skin cancer, and other health problems. One EPA study pegged the potential damage at 150 million cases of skin cancer and three million deaths during the course of the 21st century at an economic cost of $6 trillion. These facts and the solid science behind the findings of a growing ozone hole led the global community to respond. In 1985, twenty-two countries—representing most of the world's CFC production—negotiated a treaty, the Vienna Convention, to address the problem. Two years later, the same parties and twenty more countries added the Montreal Protocol to the treaty agreeing to phase out the production of CFCs. Such stringent new regulations might seem like a business burden. In Chapter 4, however, we'll see how DuPont developed CFC substitutes, then advocated tough regulations to gain a competitive advantage in a transformed marketplace.

The Montreal Protocol and follow-on amendments are perhaps the greatest success in international environmental law. Even the most morose doom-and-gloomers agree that we've made substantial progress on this problem. The ozone hole has stopped growing. If current emissions controls are maintained, it will be largely closed by about 2065.

Business Consequences

The world has banned CFCs and related chemicals. Many businesses had to find substitutes to use in the production of aerosols, solvents, coolants, and cleansers. But some of the substitutes were also found to be dangerous to the ozone layer, and they have also been banned. Some chemicals slated for phase-out by international agreements have particularly important uses and few substitutes, so controversy remains. Most notably, the U.S. government has argued that methyl bromide, which is used as a fumigant on farms, should not be banned despite its recognized harmful effect on the ozone layer.

OCEANS AND FISHERIES

Overview

The capacity of the world's oceans once seemed endless both as a source of fish and as a place to dump waste, but humanity has outstripped even this vast scale. As Unilever found out, fish stocks have decreased dramatically. Over three-quarters of the world's fisheries are over-exploited and beyond the point of sustainability. Put simply, we are catching fish faster than they can reproduce. Gigantic fishing nets scour the ocean and sweep up every swimming thing with a shocking efficiency, while traditional trawlers now deploy sophisticated sensors to find schools that previously evaded traditional fishermen. Fish have nowhere to hide.

Ocean habitats are also in trouble. About 20 percent of the world's coral reefs are dead, and more are dangerously degraded. In the end, though, our waters are really indicators of other environmental problems. Climate change kills the coral, air pollution settles into waterways, and farm runoff moves down rivers to the ocean. In the Gulf of Mexico, we've created a dead zone with virtually no sea life. At the mouth of the Mississippi River, chemicals and fertilizers have killed everything in an area bigger than New Jersey.

Business Consequences

Is the demise of our oceans a problem for most businesses? Probably not directly. But for those whose livelihoods depend on fishing, recreation, and tourism, the effect of declining fisheries may be severe. And for anyone who eats fish—a major source of protein for many

NORTH ATLANTIC FISHERY COLLAPSE

During the 1990s, North Atlantic commercial fish populations of cod, haddock, and flounder fell 95 percent. The value of the British catch alone declined $300 million from 1994 to 2002. The U.N.'s Food and Agricultural Organization estimated that sustainably managed fisheries would yield $16 billion in higher revenues.

people around the world—the issue will be immediate. As noted earlier, Unilever's response is to work to increase the supply of certified fisheries that protect populations. Other solutions are emerging, too, like increased fish farming.

DEFORESTATION

Overview

Deforestation should not be a big issue in the United States; forest cover is increasing across most of the country. But how trees are cut remains an issue. Clear-cutting scars the landscape and leads to soil erosion and water pollution. Cutting down "old growth" forest destroys precious habitat and often inspires an uproar of protest. And while some North American and European timber companies have gotten religion on the need to manage their forests with care, others around the world have not.

In South America and in some Asian countries (Indonesia, for one), deforestation is barely slowing. Logging for wood is only part of the problem. A bigger issue is converting lands (a nice way of saying cutting and burning) to agricultural use to feed a growing population coupled with a growing demand for meat that requires vast areas for cattle grazing. Even with reforestation, we lose millions of acres of forest every year. Since 1990, the net result is the destruction of forests equal in area to Texas, California, and New York combined, or in European dimensions, an area larger than Spain and France combined.

Business Consequences

Every company that uses wood, paper, or even cardboard packaging has some stake in, and responsibility for, the state of our forests. When McDonald's first realized 15 years ago that litter was an issue, it began working with a New York–based NGO, Environmental Defense, on reducing packaging. Now activists are turning on less obvious users. Catalogers are being held responsible for something that they never thought about before. Limited Brands, owner of Victoria's

Secret, has faced a torrent of protests about where its catalog paper comes from—a campaign abetted by clever ads claiming the company has a "dirty secret."

For years, service businesses with no smokestacks or other visible effects on the environment worried little about environmental issues. No longer. We'll see later how a group of big companies—some obvious (Staples), some more surprising (Bank of America and Toyota)—have banded together to deal with issues such as paper use.

OTHER ISSUES TO LOOK OUT FOR

Environmental issues resist categorization. Specific concerns are vital to some companies or industries and all but invisible to others. Every company needs to be on the lookout for fast-changing realities and how they affect its own circumstances. By way of an all-purpose checklist, here are a few additional issues to watch:

Food Safety

Fears about bioterrorism and the security of food and water supplies loom large in some people's minds. Consumer pressure to reduce chemical use in agriculture is also growing fast. The demand for organic food has skyrocketed. But perhaps the most potentially explosive issue is the use of genetically modified organisms. Many consumers, particularly in Europe, fear that genetic modification might make food more dangerous or that we'll face problems we can't predict as we "play God." To date, there is little evidence that genetically modified food causes health problems. As we'll get to in later chapters, however, we live in a world where what stakeholders feel can be just as important as the facts.

Radiation

Emissions-free nuclear power could be an answer to climate change, although serious fears remain about exposure to radiation from accidents or from improper handling of spent fuel. Many people are also questioning the use of radiation as a sterilizer. Irradiating food, for example, can be an effective way to prevent spoilage, but the word "radiation" in the same breath as "food" makes many people uncomfortable.

Desertification

In a number of places around the world, deserts are growing and encroaching on human livelihoods. Poor land use choices, overdevelopment, and the general warming and drying of the planet are the likely causes. The effects are unpredictable but could be far reaching. In 2002, schools in Seoul, South Korea, were shut down when a dust cloud blew in from China's growing deserts 750 miles away. China's dust has even reached the United States.

A TOOL TO MANAGE THIS COMPLEXITY: ISSUE SPOTTING THROUGH *AUDIO* ANALYSIS

The sheer range of environmental issues companies need to track can seem overwhelming. Before launching into the process of creating a corporate environmental strategy, every business needs to develop an "issue map." To get started, we suggest undertaking what we call an AUDIO analysis.

The purpose of this analysis is to help you "listen" to your business, up and down the value chain, for issues and opportunities. It starts with a grid of the ten key challenges that we've identified on one axis, and five categories on the other: Aspects, Upstream, Downstream, Issues, and Opportunities (see the table on pages 62–63 for a sample AUDIO analysis for a big-box retailer).

For this exercise, bring together people from a range of business functions including environmental professionals, purchasing agents, marketing executives, and so on. Together, brainstorm on what **aspects** of environmental issues touch the business. For example, do you use a great deal of energy and produce greenhouse gases in large quantities? Do you rely on fresh water? Next, look both back in the value chain (**upstream**) and forward (**downstream**) and ask a similar set of questions. How do these issues affect your suppliers or your customers? Even if an issue is not central to your business, it could be critical to your suppliers, which ultimately makes it your issue, too. You may not use toxics or heavy metals in your production, but as Sony found out, if your suppliers do, it can wreak havoc on your business. Downstream effects are also your problem. What do your customers do with your product, especially when they are finished with it? The fact that your own business does not directly cause harm will not shield you from responsibility. Your customers' problem is your problem.

After this first big picture view from 30,000 feet, dig down and ask what kind of challenges or **issues** arise from these environmental problems—both for you and the rest of the value chain. Are there areas in your operations that rely on a particular resource (like water), and what would happen if that resource were in short supply? Finally, look for **opportunities** to profit. Would lowering your product's energy use help your customers? Could you sell more product?

We'll come back to this tool in Chapter 11, but for now, think of the AUDIO as a starting point and a foundation on which to build a corporate environmental strategy. This issue map looks at how environmental dilemmas affect your business, your suppliers, and your customers simultaneously. The problems are real and here to stay. An AUDIO analysis can help you think carefully about how to develop strategies to deal with the risks and exploit opportunities presented by these issues. Only then can you create Eco-Advantage.

Simplified Sample AUDIO Analysis for a Big-Box Retailer

Challenge	Aspects	Upstream	Downstream	Issues	Opportunities
1. Climate Change	Emissions from energy use	Distribution system emissions; Supplier emissions from operations	Emissions from customers driving to store; Energy use from products sold in stores	Possible carbon constraints or charges	Launch eco-efficiency effort targeted at energy use and related greenhouse gas emissions
2. Energy	Energy consumption and rising costs	Supplier choice of energy and sensitivity to changes in energy costs	Energy use from products sold in stores	Energy sourcing and cost burden; Reliance on grid	Reduce energy use through store retrofit; Negotiate favorable rates; Sell energy-efficient products
3. Water	Contaminated runoff from buildings and parking lots	Water use in agriculture for food sold in store	Toxic products (such as lawn and garden supplies) ending up in waterways	Rising pressure for improved water quality	Redesign parking lots and runoff flows
4. Biodiversity	Habitat fragmentation due to land use and "footprint" of stores	Products that rely on or reduce biodiversity	Customer product use (or misuse) that causes ecological damage	Local development constraints and concern about "sprawl"	Invest in land conservation and build corporate reputation
5. Chemicals/ Toxics	Use of chemicals in store operations	Fertilizer runoff from farms for food sold	Chemicals in products sold	Possible liability for chemicals in environment or humans	Sell organic foods and other green products

Challenge	Aspects	Upstream	Downstream	Issues	Opportunities
6. **Air Pollution**	Air emissions from facilities	Emissions at supplier factories and energy sources	Emissions from products	Tightening controls on air emissions	Increase efficiency to reduce emissions and costs
7. **Waste Management**	Quantity of garbage generated	Solid and toxic waste from supplier production	Disposal of packaging by customers	Rising cost of waste disposal and increased "take-back" legislation	Reduce packaging and offer take-back and recycling options
8. **Ozone Layer**	Residual CFC use in refrigeration	CFC releases by suppliers	CFC leaks from products	Legal constraints on use of CFCs	Partner to develop non-CFC products
9. **Oceans**		Declining global fish stocks and rising prices		Rising costs and growing pressures to track seafood sources	Sell sustainably caught seafood
10. **Deforestation**	Land clearing for facilities	Wood supplier reliance on unsustainable timber sources		Risk of consumer protest or even boycotts	Set sourcing criteria for suppliers

THE ECO-ADVANTAGE BOTTOM LINE

Environmental problems can't be ignored. Yes, conditions are improving in some cases. But rising threats are sure to change traditional ways of doing business and even our way of life. The range of issues is daunting. A few, like climate change and water concerns, will be major problems for all of society. Nearly every company, large and small, will be forced to address them. Other problems, such as toxic exposures, hit some industries much harder than others. In the face of these diverse pressures, companies need to stay on top of the big issues, understand where the science stands, and know where in their value chain the impacts lie. The precise effects of these sizable forces on markets, industries, and companies are unpredictable, creating both risks and opportunities. But smart companies develop tools to make sense of the rapidly evolving market conditions they face.

Chapter 3 Who's Behind the Green Wave?

In 1995, activists from the environmental group Greenpeace boarded Shell's defunct North Sea oil platform, the Brent Spar. Protesting Shell's plan to dump the rig in the North Atlantic Ocean, they unfurled giant banners claiming the platform would pollute the ocean with thousands of pounds of toxic chemicals. The situation went from bad to horrendous when Shell turned water cannons on the protestors in one of the worst public relations moves in corporate history. Customers across Europe were soon cutting up Shell credit cards and boycotting its gas stations.

The irony of this international incident is that Shell's plan to sink the Brent Spar had real scientific merit. Outside experts and Shell scientists had studied the disposal plan carefully, and it was even backed by the British government. Greenpeace later admitted it had its facts wrong, exaggerating the level of pollution a thousandfold. But being right was not the point. In the court of public opinion, Greenpeace made its case and won. Its pitch was convincing and superficially plausible—after all, how could sinking a 300-foot oil platform *not* be harmful to the environment?

To its credit, Shell learned from the fiasco and is now a global leader in the evolving art of stakeholder relations, including the craft of dealing with tenacious NGOs like Greenpeace. But a company that worries only about activists and associated public relations nightmares will miss the bigger picture.

Like it or not, an ever-growing chorus is demanding that companies explain and justify how they treat the environment. Among the major new players:

- Consumers who wonder what's in the products they buy and how safe they are for themselves, their children, and the environment
- Business-to-business customers who demand that suppliers reveal how they make their products and *exactly* what's in them
- Employees who want to align their personal and professional values and need to know what their company stands for
- Banks which reinforce all these concerns by factoring environmental variables into their loan decisions
- Insurance companies that have come to view environmental risks as business threats
- Stock market analysts who study environmental performance as a signal of over-arching management quality.

All these pressures can dramatically affect a company's fortunes and determine what projects get financed, whether the best employees stay with the company, and how easy it is to get products to market. Smart companies are dealing with these pressures head on. In the aftermath of its Brent Spar embarrassment, Shell launched a well publicized "Tell Shell" campaign and now does a great deal of stakeholder management to avoid problems down the road. At its Athabasca Sands operations in Canada, the company has spent millions on countless meetings with local communities, regional governments, and indigenous populations. The goal is to make sure that everyone who can seriously affect Shell operations is heard early and fully.

Companies like Shell work hard to catch the Green Wave before it catches them. But before we dig into the how, it's vital to understand the who. Below is the Eco-Advantage "playing field"—five core categories of stakeholders who care about the environment:

- **Rule-Makers and Watchdogs** such as government regulators and environmental groups
- **Idea Generators and Opinion Leaders** including think tanks and academics
- **Business Partners and Competitors** as well as suppliers and B2B customers
- **Consumers and Community** including local officials and the general public
- **Investors and Risk Assessors** such as stock market analysts and bankers

An executive responsible for corporate social responsibility (CSR) at a large consumer products company recently asked Andrew, "Can

Eco-Advantage Playing Field

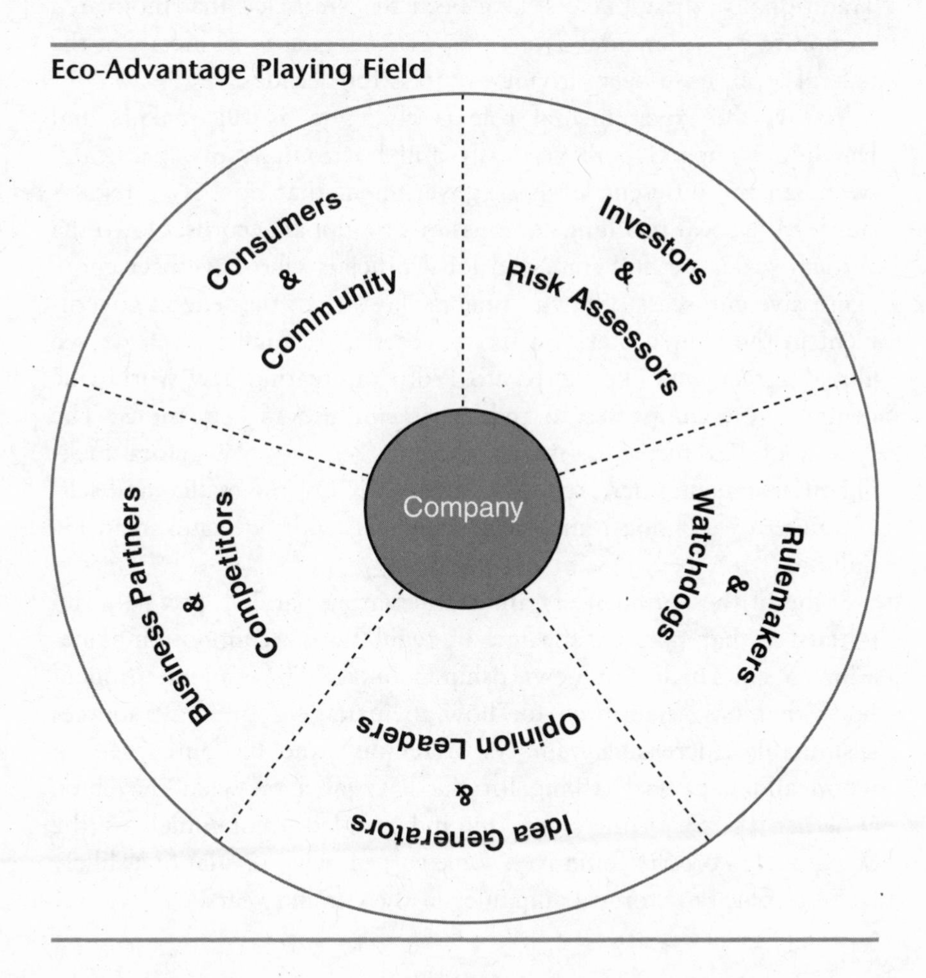

you think of a company that's embraced CSR *without* feeling pressure from outsiders first?" She stumped us—not many come to mind. Sure, a small number of WaveRiders have deep-seated beliefs and the culture to back them up. Herman Miller, the Michigan-based office furniture company, is one. But in general, big brand-centric companies don't have an epiphany without a push from true believers.

In the following brief tour through the five core stakeholder groups, we'll look at examples of how they influence companies today and at what trends will shape tomorrow's playing field.

RULE-MAKERS AND WATCHDOGS

Traditionally, these players have been the "muscle" that motivated corporate environmental awareness. Strict regulations, mostly at the federal level, have been driving progress for decades.

Today, the governmental role is changing as rule-makers and watchdogs expand both vertically and horizontally. By "vertical," we mean the different levels of government that now issue regulations, from local planning commissions to global accords. Down the vertical scale, we find state and local officials who have been more aggressive enforcers of environmental laws than the federal government in the United States in recent years. At a higher level, we see global agreements like the Kyoto Protocol creating new worldwide demands on companies to reduce greenhouse gas emissions. The "horizontal" dimension refers to the emergence of new actors tracking environmental performance, such as NGOs, the media, and self-appointed watchdogs, including bloggers with websites read by millions.

Some of these influencers are even creating parallel, privately run initiatives that take on the feel of regulations as more companies adopt them. The Forest Stewardship Council (FSC) is an independent body that has set criteria for how to manage a forest's resources sustainably. Increasing numbers of retailers and big purchasers of wood and paper are asking for the FSC label on what they buy. Whether it's FSC or the competing industry-led set of guidelines (the Sustainable Forestry Initiative), some sort of quasi-regulation is likely to be in place for forest companies in the coming years.

NGOs

It's nearly impossible to know how many NGOs are out there for sure, but one study found over 20,000 multinational NGOs. Hundreds of thousands more are based in individual countries. Many thousands at both levels are focused at least partially on the environment.

The biggest environmental NGOs have established themselves as multinational entities of immense reach and influence. These include Environmental Defense, World Wildlife Fund, Natural Resources Defense Council, Sierra Club, Greenpeace, Conservation International, Nature Conservancy, National Wildlife Federation, and Friends of the Earth. These are the old guard, with many pushing past the ripe old age of 30. Despite recent cracks in the foundation of the environmental community, these groups retain considerable public influence. One study showed that 55 percent of opinion leaders trusted NGOs, while only 6 percent trusted business.

Many of these organizations were founded on models of confrontation, whether making waves in the courts (like the Natural Re-

NEW AND SURPRISING PARTNERSHIPS

How NGOs and business interact has changed dramatically over the last few decades. All of the mainstream environmental NGOs now include engagement with the business world as a core function. They've learned to use the stick and the carrot. As the *Financial Times* noted:

> Environmentalists have smartened up their act. Gone are the woolly jumpers and sandals. In their place are smart suits and ties. . . . just days after Greenpeace activists scaled a British Petroleum offshore rig to protest against new oil exploration . . . [Greenpeace director] Mr. Melchett had dinner with John Browne, British Petroleum's chief executive.

NGOs have only gotten more sophisticated as interactions have expanded from the occasional dinner to full-fledged partnerships. At the 2002 Earth Summit in Johannesburg, Greenpeace stood shoulder to shoulder with the World Business Council for Sustainable Development to seek global action on climate change.

sources Defense Council) or sitting in trees and boarding oil platforms (like Greenpeace). Pressure can be global as in Shell's case, or it can shine like a spotlight on key executives at corporate headquarters. A few years ago, some clever agitators stood outside Fenway Park in Boston to hand out baseball cards depicting both the Chairman and CEO of Staples, Inc. as perpetrators of global deforestation. Printed on the back of the cards were some strong criticisms of the office supply giant. In the words of one Staples executive, "They basically accused us of single-handedly clear-cutting the world's forests."

Sometimes these "in your face" tactics can go too far, such as when Michael Dell's wife faced angry protestors outside her place of business about Dell's lack of electronics recycling. Or they can go way over the line to illegal and dangerous action. In 1998, Vail, the famous Colorado ski resort, was the victim of arson when extremists torched multiple buildings, supposedly in protest over the resort's expansion into pristine habitat. No company can plan completely for irrational action, but every company can greatly reduce the odds of even coming under attack. Other ski areas—most notably Aspen—have avoided aggressive criticism by making environmental concerns a central element of their business strategies. In fact, the Aspen Skiing Company recently committed to using wind power for 100 percent of its operations. Respected watchdog groups rate the company the number one ski resort in the West on environmental dimensions, while Vail continues to get low marks.

In fact, confrontational approaches are on the decline. NGOs are now choosing to partner with companies at least as often as they attack them. In a later chapter, we further explore partnerships and how to make sure they are effective and create real value, but we want to highlight here two of the longest running and best NGO–corporate partnerships out there:

- **Chiquita and the Rainforest Alliance.** A company with a checkered past, Chiquita felt immense pressure to change its ways in the early 1990s. Over the next decade, working closely with Rainforest Alliance, the company transformed its Latin American banana operations.

- **McDonald's and Environmental Defense.** McDonald's may be the king of fruitful NGO partnerships. The company started in the early 1990s working with Environmental Defense on packaging, and together they eliminated the Styrofoam "clamshell" in which McDonald's had sold a billion-plus hamburgers.

Even though NGOs frequently collaborate with companies these days, the watchdog role remains very much alive within many groups. Often an NGO can work with a company on one issue and attack it publicly on another. Companies find this more than a little frustrating, but it's a reality they must face. Whether global in focus or local in scale, whether pushing their agenda from behind the scenes or in front of the cameras, whether making nice or making trouble, NGOs are a force to be reckoned with. A company without a plan or strategy for dealing with them runs a growing risk of getting dunked.

Government and Regulators

Jim Rogers, the CEO of energy producer Cinergy, has said that he's always worried about new regulations that can change the value of a business instantly—what he calls "stroke of the pen" risk. The world of environmental regulation *is* dynamic. Legislators are constantly revising federal laws and agencies routinely issue new regulations. More dramatically, regulatory requirements are emerging from new places in the legal hierarchy. From ordinances issued by local zoning commissions to directives from the European Union, environmental law is getting ever more complicated and multilayered.

Companies that fail to track regulatory developments risk serious competitive disadvantage.

"VERTICAL" CHANGES—LOCAL PRESSURE

Across the United States, local and state governments are stepping in to fill gaps they see in environmental management. About thirty U.S. states have plans to cut greenhouse gases within their borders, often including renewable energy portfolio standards that force utilities to generate up to 25 percent of their energy from non–fossil fuel sources. California has adopted tough new air quality standards requiring a 30 percent reduction in greenhouse gas emissions from vehicles by 2016. This mandate promises to reshape the market for cars and trucks. Auto companies who are ready with "clean car" technologies, such as hybrid gas–electric engines, will thrive. Others will suffer lost market share and decreased profitability.

Even cities are getting in on the act. Mayors across the world are taking action to reduce greenhouse gas emissions, cut back on waste, and control sprawl (see box).

MAYORS ON THE MOVE

In early 2005, the Kyoto Protocol, an international agreement to cut greenhouse gas emissions, went into effect. It was ratified by 160 countries—with the notable exception of the United States. Seeing that gap, Seattle mayor Greg Nickels led the U.S. Conference of Mayors to unanimously adopt its own Climate Protection Agreement. Over 350 mayors of large and small cities from across the political spectrum have individually committed to match Kyoto goals and cut greenhouse gas emissions.

"VERTICAL" CHANGES—INTERNATIONAL PRESSURE

At the other end of the vertical spectrum, new international regulations promise to reshape the competitive playing field. In the face of global-scale climate change concerns, China has implemented stringent requirements for fuel efficiency in cars that are five miles per gallon *higher* than U.S. standards. Guess which superpower's auto companies are least prepared to meet these emerging market standards.

In recent years, the European Union has developed a series of hard-hitting directives. Some observers believe these new laws will greatly improve environmental quality, while others say they will destroy entire industries. They will almost certainly do both. Three directives in particular are having enormous impact:

1. Restriction of Hazardous Substances Directive (RoHS, pronounced "rose")
2. Waste Electrical and Electronic Equipment Directive (WEEE)
3. Directive on Registration, Evaluation, and Authorization of Chemicals (REACH).

The first two focus mainly on toxics and recycling in the electronics industry, but they are powerful indicators of the laws coming down the pike. Today, the focus is on computer and cell phone manufacturers. Tomorrow, many more industries will be on the hot seat. And don't think these rules affect only European companies—any company making a product for the EU market must comply.

RoHS bans a list of hazardous substances such as lead, mercury, and cadmium from use in new electrical equipment. Manufacturers have been working hard to find substitutes for the lead solder that holds integrated circuits together. Of course, there's been grumbling about the selective focus of the requirements and questions raised about how effective RoHS really will be. But perfectly constructed or not, this directive is not going away. The rule also represents an interesting shift in regulatory strategy. As IBM's top environmental executive Wayne Balta puts it, "Laws like RoHS show the movement from regulating outputs to regulating inputs—from what comes out of the smokestack to what goes into products."

WEEE concerns the other end of the product's life cycle. It demands that all manufacturers from a wide range of industries, from electronics to appliances, pay for proper disposal or recycling of their products. WEEE represents the latest development in the growing category of take-back laws that place cradle-to-grave responsibility for products on the manufacturers. These laws encourage value chain thinking by imposing a real cost on companies that do not design products with the end of life in mind. Some estimate that RoHS and WEEE compliance could translate into a surcharge of as much as three percent of the cost of goods sold—not a trivial amount.

Forward-thinking electronics producers are starting to explore designs that eliminate all toxic substances so they'll be ready for any

THE POLLUTER PAYS PRINCIPLE

The long-term regulatory trend in every country we've studied is toward making those who pollute (1) limit their harms and/or (2) pay for the damage they cause. This Polluter Pays Principle ensures economic efficiency and protects property rights, giving it a strong legal logic as well. The push for companies to "internalize costs," as economists would say, means that a competitive edge based on sending pollution up the smokestack or out a waste pipe will be hard to maintain over time.

possible legislation. Even if no future legislation comes, the effect of these directives on the electronics industry has already been profound. The American Electronics Association declared that the new rules "fundamentally alter every high-tech company's business strategy not just for the European Union, but also for its global supply chain management."

When we spoke with executives at Intel, Dell, and AMD, we heard repeatedly that the laws are forcing changes in their designs and processes globally. It makes no fiscal sense, they say, to produce chips or assemble electronics one way for distribution in a market as big as the European Union and differently everywhere else. Bottom line: The EU's aggressive policies are driving change around the world.

OVER-REACH-ING?

The most ambitious piece of legislation working its way through EU headquarters in Brussels is the Directive on Registration, Evaluation, and Authorization of Chemicals, known as REACH. This sweeping restructuring of Europe's approach to the regulation of chemicals threatens to realign a number of major industries.

REACH requires producers to register every chemical they make—about 30,000 in all—and measure the potential risk to public health. REACH builds on the idea that society should not introduce new materials, products, or technologies if the risks are unknown. It's the poster child for the precautionary principle. The directive shifts the burden of proving safety from governments to business. Companies will have to demonstrate that a product is safe or that the benefits to society outweigh the risk.

Many companies, including several pharmaceutical giants, claim

REACH will kill innovation and become a competitive drag on companies operating in Europe (see box). Advocates for REACH say the requirement that companies prove the safety of their products is long overdue and will reduce the risks of releasing chemicals with unknown affects on humanity. Both sides make plausible arguments—REACH could be needed for public health *and* compliance just might cost billions of euros while constraining new product development. Either way, REACH will undoubtedly have a large impact on some very big industries.

BEATING THE COMPETITION ON REGULATORY COMPLIANCE

Business leaders often overestimate the cost of regulations and underestimate their own capacity for innovation. For example, industry estimates of the compliance costs for cutting acid rain under the U.S. Clean Air Act of 1990 ranged up to $1,500 per ton. Over the first ten years of the program, the price per ton never went above $200, and was usually far lower. As BP's Lord John Browne has said about his industry, "Every time there's a new piece of legislation, we say it's the end of our industry. . . . [We have] an appalling track record in this regard." Companies that figure out cost-effective ways to comply with regulations can cut costs relative to their competitors and establish Eco-Advantage.

BEYOND REGULATION: A NEW GOVERNMENTAL TOOLBOX

Over time, governments around the world have broadened their approaches to environmental protection. The French, for instance, have recently enshrined the right to a safe and healthy environment in their constitution and deeply embedded the precautionary principle in their regulatory regime. While some use of "command and control" mandates persists, the shape of regulation is changing.

Rather than dictating a particular pollution control technology, governments are coming to recognize the value of setting performance standards, which give companies discretion in how to respond. Many countries, led by the United States, have moved toward regulations that use economic incentives. Such market mechanisms include taxes on pollution or polluting products, tradable emissions allowances, and deposit-refund schemes for harmful products such as batteries. The United States has used market mechanisms to phase out lead from gasoline, eliminate ozone layer–damaging CFCs, and reduce acid rain.

Europe is using an emissions trading market to reduce greenhouse gas emissions and meet its requirements under the Kyoto Protocol. Under this scheme, thousands of facilities (utilities, refineries, and so on) across the continent receive greenhouse gas allowances. The companies must then either reduce their emissions to meet their quotas or buy more pollution allowances in the marketplace.

Even well-intentioned regulatory regimes that harness economic incentives and promote maximum efficiency can have negative effects on business if not carefully structured and coordinated. The EU's Kyoto Protocol rule, adding charges of roughly $30 per ton on carbon dioxide releases, promises to impose significant burdens on companies and even whole industries. If the results were simply to reward those who are most effective at cutting greenhouse gas emissions, the reshaping of markets might be of little consequence. But this pollution surcharge could broadly diminish Europe's industrial competitiveness unless companies find low-cost emission reductions or until major competitors, including the United States and China, take on the same burdens.

Another important trend is the move toward "information regulation." The Toxics Release Inventory in the United States requires companies to report on their chemical releases to air, water, and land. Similarly, the PROPER law in Indonesia gives industrial facilities public, color-coded scores based on their pollution management. The ratings run from black for no pollution control to green for significant pollution prevention efforts to the top score of, yes, gold for facilities that keep emission levels near zero.

The pressure for transparency focuses on more than reporting on emissions and pollution. Companies are increasingly required by the Securities and Exchange Commission, the Financial Accounting Standards Board, and the European Union to share information about their environmental performance and how it affects their financial status. Although implementation rules are still being developed, the United States' Sarbanes-Oxley Act (see box) appears likely to require companies to disclose much more about environmental risks than ever before. And the law makes CEOs responsible for knowing about these liabilities or risk going to jail for their ignorance.

SARBANES-OXLEY ACT OF 2002

As often happens in the U.S. Congress, quick reactions make for strange legislation. In 2002, legislators felt pressure to do something about the scandals plaguing the corporate world. One significant result was the Sarbanes-Oxley Act, a law aimed mainly at improving financial liability disclosure but causing a flurry of activity on related, broader topics. A critical clause of the act, Section 401(a)(j), mandates that companies "shall disclose all material off-balance sheet transactions, arrangements, obligations . . . that may have a material current or future effect on financial condition." The "material" issues in this broad provision, and the related Section 409, include environmental liabilities.

The precise contours of what Sarbanes-Oxley requires are still being clarified. Some potential liabilities are fairly obvious, like the costs of cleaning up a contaminated manufacturing site. But what about more remote risks? Should companies estimate their potential liability for contributing to climate change? The only certainty is that the trend is toward more disclosure, not less.

Disclosure as a category includes more than just financial risks. A growing body of "right-to-know" laws requires that companies disclose potential risks to public welfare. The U.S. Toxics Release Inventory is the best known law of this sort, and the gold standard. But other, smaller laws are making waves as well. California's Proposition 65 forces companies to label products containing any chemical that creates a 1-in-100,000 risk of cancer. Faced with having to label their products as carcinogenic, tuna packers removed lead from the solder used to seal cans. And winemakers have shifted from lead foil to plastic bottle-top wrapping.

While some disclosure laws mandate the release of information on negative environmental attributes, others focus on highlighting positive characteristics. The U.S. EPA has created a number of programs and awards, such as Energy Star and Climate Leaders, to recognize companies with top-tier performance. Governments are also using substantial purchasing power—buying everything from recycled paper to buses powered by natural gas—to make markets in environmentally preferable products.

Don't believe all the talk about governments being stodgy bureaucracies. Sure, some officials do seem to think they are in the business of producing red tape, but many more in city halls, state capitals, and Washington (or Brussels) are actively seeking new and clever ways to reduce emissions, save open space, and conserve the land's natural bounty. Smart companies do not see regulators as enemies. Instead, they work with government officials to shape incentives and create successful environmental programs. It makes much better strategic sense to partner with regulators and anticipate their demands (and public expectations) than to develop contentious relationships and be unwillingly prodded from behind.

Politicians

Dedicated regulators rarely play to the camera, but politicians are always running for reelection. To show how they stand up for the wronged little guy, they'll publicly rebuke suspected bad actors.

Political showboating is endemic all over the world. Take an example from the Philippines. After floods and landslides killed hundreds of people, President Gloria Arroyo blamed illegal loggers. The logic of her criticism was that forests retain water, lessen the effects of floods, and help reduce landslides. Since the line between illegal and legal logging is not very distinct, the forestry industry in general looked pretty bad. Arroyo's criticism was not misplaced. Bad forestry practices and clear-cutting do make floods worse. But blaming the effects of natural disasters solely on one industry is classic political theater.

To avoid becoming a politician's scapegoat, we suggest concentrating on a few key points. First, build relationships with elected officials up and down the political ladder. Second, "immunize" your company by building a deserved reputation for environmental stewardship. Finally, and most fundamentally, don't present an easy target through bad behavior. In the political arena, big business and those whose behavior can be portrayed as outrageous on any dimension stand out.

The Plaintiff's Bar

One of the biggest environmental dangers every company faces, especially in the United States, is the risk of being sued for pollution or ecological harm. No company can afford to ignore the "plaintiff's bar," lawyers who specialize in bringing such tort law claims.

From asbestos class action law suits to neighbors claiming that a factory is a nuisance, legal action is not going away anytime soon. If anyone thinks the travails of the asbestos industry represented an isolated problem, think again. In 2006, a landmark case in Rhode Island found producers of lead paint legally responsible for the negative health effects of their products. Legal threats are growing and getting more sophisticated. One company's top environmental official recently told us that she considers the plaintiff's bar to be *the* stakeholder to watch.

A few brief additional examples of interesting cases that multinational companies need to take to heart:

- W.R. Grace, which once had a multibillion dollar market capitalization, was driven into bankruptcy in 2001 by an avalanche of asbestos-related lawsuits.
- Plans for a new coal plant in Australia were scuttled by a lawsuit focusing on greenhouse gas emissions and the facility's contribution to global warming.
- A class action lawsuit for Hurricane Katrina victims was filed against ten oil and gas companies for destroying wetlands that might have reduced the severity of the floods.

"Frivolous. Preposterous. It won't happen to me." So you might say. But beware the plaintiff's bar. It's stocked with creative, passionate lawyers looking for new lines of attack.

Don't miss this vital point: Even if a company fully complies with the law and ultimately wins in court, it may be hurt badly by the fight. Defending against lawsuits can be prohibitively expensive. Even when a company wins, unflattering facts can emerge that will doom it in the court of public opinion. That's why going beyond compliance and managing stakeholder relations carefully is so vital.

IDEA GENERATORS AND OPINION LEADERS

Media

From the Love Canal hazardous waste crisis, which led to the U.S. Superfund program, to the *Exxon Valdez* oil spill, which triggered the U.S. Oil Pollution Act of 1990, to hundreds of smaller, local pollution stories, media reaction drives public understanding and shapes the political response. To turn green into gold, you need to manage media relations with care, but that's not as easy as it used to be. Beyond TV, radio, and newspapers, the rise of the Internet means that the "media" has become utterly diffuse. Anyone with a video camera, a website, and an opinion can break a story.

Most dramatically, bloggers, self-appointed on-line commentators, are reshaping how news stories unfold. Blogs, both anonymous and open, are popping up from *within* companies with frank discussion about the company's products, executives, policies, and actions.

How does a company handle this? "Do nothing that might embarrass you" is the easy answer, but not especially helpful. Better to avoid presenting a prominent target. This means not only managing environmental issues systematically and earnestly but also looking for and eliminating vulnerabilities anywhere in the product's entire life cycle. Are your suppliers dumping toxic waste in a river in the developing world? Do your customers dispose of your product in a way that creates an environmental problem?

Every company needs an emergency response plan. When an issue emerges or an accident occurs, the executive team can't wing it with the public and the media. Exxon's plodding response to the 1989 *Exxon Valdez* oil spill in Alaska sealed the company's reputation for environmental insensitivity. To this day, Exxon and the *Valdez* serve as poignant reminders of corporate malfeasance, regardless of whether the company warrants the distinction. The key to positive media coverage—or the least negative possible under the circumstances—is real action to mitigate the harm. Turning crisis management over to a public relations firm, as Exxon did, is a surefire way to add fuel to the media flames.

Think Tanks and Research Centers

The media can spread or kill ideas, depending on the coverage. But reporters get their leads from somewhere. Over the last few decades, think tanks have provided many of the ideas that have framed the public policy agenda and shaped the political debate. In very fundamental ways, a few key think tanks, such as the Heritage Foundation, American Enterprise Institute, and Cato Institute, have realigned public attitudes about the role of government in society.

In the environmental realm, several groups have played a similar transformational role. Resources for the Future led the charge to shift environmental protection strategy from command and control regulation to the use of market mechanisms such as pollution charges and tradable emissions allowances. Another Washington-based group, the World Resources Institute, has been instrumental in sharpening the focus on the link between economic development and environmental progress and has added momentum to the concept of sustainable development.

To stay on top of the leading ideas that set the framework for environmental and social strategy, companies need to monitor these important idea generators. Strategic ties to think tanks offer a way to do this. At the very least, companies should track the policy proposals coming out of these and other leading research centers.

Academia

New ideas across the spectrum, from policy to science, also come out of our institutions of higher learning. Creating links with universities can help a company stay at the cutting edge on evolving issues. In a knowledge-based economy, it makes sense to be connected to those in the knowledge production business. In addition to the flow of ideas, companies may find that these relationships provide a mechanism for connecting with talented future employees.

More and more companies are creating these linkages. Microsoft partners with the Indian Institutes of Technology to forge greater ties with the part of the world producing the most new software engineers. BP has developed structured relationships with a number of universities and uses these links to help refine its strategic planning.

Before launching a new China initiative, top company officials paid a visit to Yale University to review with academic experts a range of issues from Chinese history to corporate governance and environmental challenges.

Not all of these links focus on pollution or natural resource management concerns. But when problems crop up—and they will—a working relationship with one or more universities or research centers gives a company a place to turn for new ideas and perspectives. In a marketplace that puts a premium on innovation and fresh thinking, connections to centers of knowledge provide a wellspring for generating Eco-Advantage.

BUSINESS PARTNERS AND COMPETITORS

Managing the traditional business playing field—supply chain, customers, and competitors—lies at the heart of any successful strategy. With environmental issues, this basic reality does not change. While NGOs get all the headlines and government regulations shape the marketplace, the traditional players continue to exert real pressure, forcing companies to realign their business plans to meet changing expectations. But smart companies don't just react. They manage these relationships proactively for strategic advantage.

Industry Associations

Environmental reputations are often defined for a whole industry. The major chemical companies know this well. When Union Carbide's facility in Bhopal, India, exploded in 1986, the entire chemical industry faced a make-or-break moment.

The industry responded with a broad-based initiative to tighten standards for chemical manufacturing, storage, and transportation. Under the banner of "Responsible Care," DuPont, Dow, and other leading chemical companies committed themselves to environmental requirements well above those set by law.

The industry association has continued aggressive self-regulation. By doing so, the major chemical companies have fended off government mandates, rebuilt the reputation of the industry, and pressured sub-par performers to improve their environmental stewardship. All association members are now required to have environmental management systems and to seek third-party certification of Responsible Care practices. Although the industry still has plenty of critics, Responsible Care has filled a void, making a higher level of safety and environmental focus mandatory for entry into the community of chemical producers.

A range of industries from forestry and coffee to apparel and electronics now use industry-wide initiatives to set guidelines for social and environmental practice. As reputations are ever more connected, industry pressure to meet acceptable minimum standards is a growing trend that will not slow down.

Beyond standards, industries often collaborate in other positive ways. Working with colleagues offers a number of benefits. It provides the comfort and security of group action; no one company sticks its neck out. Combined effort may lead to more cutting-edge science, policy, and analysis as everyone pools resources to find solutions to group problems. Finally, group members can exchange best practices, raising the bar for everyone.

Industry associations have their dark side as well. They can cover for their member companies and mask the source of opposition to new regulations. In a few cases, such associations have far exceeded the bounds of appropriate advocacy and sought to confuse the policymaking process. The now-disbanded Global Climate Coalition, a seemingly neutral name for a fossil fuel industry organization, was notorious for trying to obscure the emerging science on global warming.

As an individual company, remember that your reputation is inescapably linked to that of your industry. And the industry associations you're a part of will reflect on you for good or bad.

Competitors

Even if no industry organization is driving change, one company's leadership and bold action can change the competitive playing field, sometimes in dramatic ways. In 1990, Heinz's Star-Kist tuna seized an environmental advantage over its competition by publicly committing to fish in a manner that avoided dolphin deaths. With a "dolphin-safe" label, Star-Kist's market share climbed rapidly. Others soon had to match Star-Kist's pledge as children all over America wouldn't let their parents buy tuna from a company that hurt dolphins.

More recently, facing questions about its supply chain, the Gap released a groundbreaking 2004 Corporate Social Responsibility report providing detailed data on its suppliers' compliance with environmental and social standards around the globe. In a flash, the company accepted the principle of extended producer responsibility and continued the transformation of its industry.

As one VP at a multibillion dollar manufacturing company told us, "I know I need to do something different on the environment because competitors are doing things, but I don't know why exactly." This defensive position is a recipe for trouble. Smart companies keep a close eye on competitors. They even look to partner with the enemy on thorny issues if industry-wide solutions make sense.

Business-to-Business Buyers and Greening the Supply Chain

Big customers can be a source of great stress. They want ever lower prices with no cut in quality or service. But increasingly, they are also demanding information on what's in everything you sell them, where it comes from, and how it's made. Wal-Mart, for example, has suppliers across the world scrambling to meet its new sustainability demands. The retail giant is pushing major suppliers like Kellogg and General Mills to produce organic versions of well-known brands like Rice Krispies.

"Greening the supply chain" is the technical term for all this activity. Customer pressure represents one of the fastest emerging and most powerful forces in our long list of players. In many industries, proof of environmental responsibility has become a requirement for

GIVE ME A LEVER . . .

The ripple through supply chains is no accident. NGOs are smart enough to go after large customer-facing brands (such as Victoria's Secret and McDonald's) that can apply pressure up the supply chain. As Archimedes said, "Give me a lever long enough . . . and I shall move the world." Big buyers are the end of a very long lever, and NGOs are pushing hard.

getting major contracts and keeping customers. But the conversations can be more positive as well—like when Nike calls on DuPont to talk about how they can help create a sustainable shoe. The ripples of this trend are moving through many ponds. A couple of examples:

- In 2003, Boise Cascade, which had resisted calls to protect endangered forests, announced that it would no longer harvest from some forests in Chile, Indonesia, and Canada, and that it would stop logging U.S. virgin forests. The pressure for the change came both from NGOs and big corporate clients such as Kinko's, which had dropped Boise Cascade as a supplier.
- Limited Brands recently faced a very public campaign against its Victoria's Secret line. Forest Ethics, an activist NGO, protested the fact that the paper in millions of Victoria's Secret catalogs came from fragile forests in Canada. In response, the company asked International Paper, its big paper supplier and the company actually harvesting in that region, to develop alternative solutions.

Suppliers

We expected to find buyers insisting that their suppliers meet ever more stringent environmental standards. More surprisingly, we've also seen suppliers pressure their big customers. When Dell, Inc. felt some heat from NGOs because of paper use in its catalogs, Pat Nathan, the newly minted Sustainable Business Director, took up the issue. With Michael Dell's strong support, Pat formed a working group with the procurement people. Together, they steered the company toward 10 percent recycled paper. No surprise. But what happened next is really interesting. The company realized that while it was not one of the top ten catalog mailers (as the NGOs had suggested), many of its customers were. Big mail order companies have call centers and other facilities with many computers, often supplied by Dell.

Dell sent a short letter to its customers suggesting they explore recycled content paper. The request was positioned as part of Dell's continuing effort to serve its customers. "In addition to providing a reliable and cost effective product and service," the letter said, "it is important to share best practices in areas such as supply chain and global citizenship."

Perhaps "pressure" isn't the right word for this, but questions and suggestions from suppliers will get more forceful. For years now, product categories such as alcohol, tobacco, and firearms have seen lawsuits aimed at manufacturers for what their customers do with the products. That sensibility may be spreading. As take-back laws rise in importance, more companies will look forward in the value chain to make sure their products—and any environmental problems and liabilities associated with them—don't come back to haunt them. As they say, watch this space.

CONSUMERS AND COMMUNITY

CEO Peers

The days of corporate titans deciding the fate of markets on the 9th tee seem to be on the wane, but leadership networks remain strong and important. Corporate bigwigs meet in many different contexts, from the health club to the board room to the charity ball. And they compare stories. No executive wants to be the one who created an environmental mess. They all want to be seen as upstanding citizens. Pro-environmental speeches from CEOs like BP's Lord Browne and Wal-Mart's Lee Scott—whether or not their companies are currently at the leading edge—up the ante for all executives. Peer pressure does not end in high school.

CEOs now face a world of ever-greater transparency and increasing emphasis on performance metrics and corporate rankings. Today, indicators cover not only sales and market share but also social and environmental results. Enjoying Sunday brunch at the club gets trickier when a CEO's company is bottoming out on the latest green index.

Consumers

Companies love their customers, but they also fear them. Every few years, a new wave of books comes out praising the virtues of customer focus above all. Consumers, however, can be fickle, and nowhere more so than in the environmental arena.

Give consumers products that support vitality and a healthy lifestyle, and the younger ones especially are likely to beat a path to your door. (In the trade, this is known as the LOHAS market, which stands for Lifestyles of Health and Sustainability.) But offer them products that are demonstrably good for the environment more generally, and the results are far harder to predict. Unilever got clobbered, for instance, when it tried to introduce concentrated detergent that reduced packaging. Consumers just thought they were getting a smaller box of soap at a higher price.

Despite this note of caution about relying on a huge turnout for "eco" products, certain product areas have succeeded in turning green to gold. In the food business, sales of organic products are growing much faster than the rest of the industry. Across the United States, there isn't enough organic milk to meet demand. In the consumer products arena, companies like Tom's of Maine have found a profitable niche in designing natural alternatives for everyday personal care items like toothpaste and shampoo (so profitable in fact, that the relatively small company was recently bought by Colgate for $100 million). And with sky high gas prices, it won't be just the Toyota Prius that is selling on fuel efficiency in the coming years.

Kids, the "Future"

This may seem like an odd category, but many executives told us they want to do right by future generations. Sometimes the pressure comes from within their own households. As he was contemplating accepting the CEO's role at Gap, Paul Pressler's daughter asked, "Dad, doesn't Gap have a bunch of sweatshops?" He still took the

RELIGION AS A NEW (AND VERY OLD) INFLUENCER

In the summer of 2003, a group called the Evangelical Environmental Network launched a campaign titled "What Would Jesus Drive?" The group ran ads encouraging Christians to buy fuel-efficient cars. Increasingly, religious consumers feel that caring for God's creation is a moral imperative. A poll of U.S. evangelicals showed that 48 percent rated the environment as an "important" priority—just behind abortion at 52 percent. Richard Cizik, an executive with the National Association of Evangelicals said, "That's an amazing statistic, considering that we've been talking about abortion for thirty years and we haven't even begun to make a case to a lot of our folks about environmental issues."

In 2006, a group of evangelical ministers committed themselves to the cause of stopping global warming. This move shocked some who thought the evangelical movement was a wholly owned subsidiary of the Republican Party. In breaking with the Bush Administration, the ministers made clear that Christians have a duty to stewardship of the Earth—and they ran full-page ads in the *New York Times* and *Christianity Today* saying so.

job, but it can be no coincidence that Gap's 2004 CSR report took a giant leap forward on transparency about worker conditions.

Sometimes the pressure is more organized. The "official" story on Chiquita's historic turnaround on social and environmental issues goes something like this: sophisticated consumers and powerful food-buying co-ops in Europe demanded better performance from Chiquita—the company had no choice. Of course, customer pressure was important, but as Dave McLaughlin, Chiquita's chief environmental executive on the ground in Latin America, told us, something else played a key role. "We give Europeans lots of credit, but the biggest effect came from kids from elementary schools in the U.S." The children's magazine *Ranger Rick* had started a postcard campaign to the CEO of Chiquita. Thousands of kids made appeals, and executives felt both moved and obligated to change the company's ways.

Kids can also put pressure on governments. When the *Exxon Valdez* ran aground, kids struggled to cope with the images of birds and other wildlife drowning in oil. Schools around the country asked kids to draw pictures of what they saw and felt. As a Special Assistant to the EPA Administrator, Dan Esty had the unenviable task of crafting a response to the thousands of schoolchildren who sent in pictures of the oil-soaked beaches and dying birds of Prince William Sound.

Communities

In 1995, Alcan, a Canadian aluminum company, learned a difficult lesson about the need for community support. This was not a company that was accustomed to mismanaging community issues. When it had to close one of its oldest smelters in Scotland, Alcan handled the disruption to local workers extremely well. But in British Columbia, its relationship with local communities was practically nonexistent—at best it was a zero sum game between corporate and community needs.

Alcan had done little to integrate its business plans with local interests and NGO concerns. For years that worked just fine. But when the company tried to divert a river to create hydro power for a giant smelter, it learned that times had changed. The local indigenous population, called "the peoples," objected. Traditionally, Alcan looked to the Canadian government to handle community issues, but that wasn't working this time. To continue its water-diversion project, the company needed direct community buy-in, but it had no experience in moving the local population in its direction and no accumulated goodwill, either.

Dan Gagnier, Senior Vice President for Corporate and External Affairs, told us, "We 'won' the environmental hearings and made the changes required of us to get permits . . . but the political environment had changed, so we actually lost." Added Environmental Director Paola Kistler, "In the past, we thought government represented communities . . . but we see now that we have to directly involve stakeholders—it's a changing world." In the end, Alcan walked away from the project, leaving over *half a billion dollars* in stranded costs. The company still owns half a tunnel.

As the Alcan case shows, it's vital for companies to engage local communities before, during, and after opening or expanding operations in a region. Local concerns are a rising priority for many companies as breaking ground on new buildings or factories gets harder and harder. Given the deeply complicated issues associated with indigenous communities, anti-sprawl campaigns, and not-in-my-backyard attitudes, local support or opposition can make or break even the best-laid plans.

WaveRiders learn to develop extensive community outreach. They recognize that conversations with local leaders and groups are not just a good practice, they are now a business imperative.

INVESTORS AND RISK ASSESSORS

Employees

Employees are perhaps the most powerful of the players since they can spell defeat or victory for an initiative or an entire company. They fit every category, from watchdog to community member, but we place them here for their role as investors of their time and skills in the companies they work for. Competition for talent is getting ever more fierce, which means any edge that makes an employer more desirable is worth pursuing. Without diving into too much psychology, what employees need from a workplace is shifting dramatically, particularly in the developed world. Companies want committed employees, and employees want companies they can commit to.

In 2004, Stanford Business School surveyed graduating MBAs to gauge how important different aspects of a potential employer really were. How much money would students trade to land a job with a company in the right location or with the right values? The surprising result: 94 percent would give up some salary to work for a company that cared about employees, cared about stakeholders, and committed to sustainability. On average, these supposedly money-grubbing business students would give up $13,700 per year.

Over the last decade, CEOs have been waking up to the new dynamic. The executives of the future want a company they can brag about, not just on Wall Street, but on Main Street. Even the hardest of hard-nosed CEOs, GE's "Neutron" Jack Welch—who was more than a little skeptical about the role of environmental issues in business strategy—saw this trend coming. While he battled with regulators over the role GE played in dumping toxic chemicals in New York's Hudson River, he told his executives, "Good people won't want to work for us if we don't get on the right side of this."

Values-driven employees create values-driven companies. As our WaveRiders are finding out, infusing a company with larger principles can significantly improve morale and commitment. Doing so can even save the company. CEO Anne Mulcahy has seen Xerox through dark times and near bankruptcy, and she believes that the company's commitment to corporate social responsibility saved it. "At the depth of our problems, we asked employees to roll up their sleeves," she says, "and most stayed with the company because they believed in what the company stood for . . . that we were 'grounded' in being a good corporate citizen."

So, yes, managing all these players well is about more than downside risk reduction and cost control. It offers potential upside benefits in increased productivity, lower turnover, and inspired employees.

Shareholders

Traditionally, we think of shareholders as being narrowly focused on corporate profits, not good citizenship. In fact, this simple logic is rapidly breaking down. "Shareholders" are hardly one monolithic group. True, average retail shareholders know very little about sustainability, but more and more people are investing through mutual funds and other vehicles that screen companies for social or environmental responsibility.

These investments fall under the category of socially responsible investing. According to the nonprofit Social Investment Forum, over $2 trillion in assets are screened in some way. That number is a bit misleading since it includes any fund that avoids "sin" stocks like tobacco or gambling. The assets of funds that really look for the best environmental or socially responsible companies to invest in are more likely around $200 billion—hardly chicken scratch, but dollars tell only part of this story.

Even Wall Street, never a hotbed of environmental thinking, is feeling the effects of the Green Wave. A number of stock pickers now see environmental management as an indicator of good general management. As resource limitations—especially related to fossil fuels—increasingly impinge on corporate performance, more investors will include a company's environmental strategy as a variable in their analysis. In fact, a recent Merrill Lynch report picked auto stocks based on which companies were prepared to deal with a "world of finite resources" and the "clean car revolution." No surprise, Toyota was a winner, along with Hyundai and the parts manufacturer BorgWarner.

The infrastructure is growing for identifying companies that are more environmentally responsible. The Dow Jones Sustainability Index and FTSE4Good in Europe steer investors—and companies looking for a benchmark—toward superior performers. These lists are backed by research from such firms as Ethical Investment Research Service, Innovest, and Sustainable Asset Management. They assemble data on companies' environmental strategies and outcomes, then rank or "tier" the companies. Mirroring the debt rating system, Innovest ranks companies from AAA down to CCC.

Niall Fitzgerald, Chairman of Reuters (and former Chairman and Co-CEO of Unilever), believes these rankings will matter more and more: "In time people will take the FTSE4Good seriously. It won't be for soft, social reasons, though. It will be because people will understand that if you don't operate responsibly wherever you are, your ability to operate in those places will diminish."

Even when Wall Street firms are not clamoring for better environmental results, other elements of the capital markets are pressuring companies. A series of institutional investors, in particular pension funds, are jumping on the bandwagon. In May of 2005, state treasurers and comptrollers representing trillions of dollars of investment clout met at the United Nations to discuss environmental risks. Insurance companies and union pension investors were also on hand, as the Associated Press put it, "to debate ways to pressure more U.S. companies into openly acknowledging the financial risk of climate change and exploring ways to reduce it."

Some funds are going further. Two of the three largest pension funds in the United States, both in California, set aside more than a billion dollars for direct investment in environmental companies and technologies. California Treasurer Phil Angelides took heat for betting state pension funds on the growth of environmental markets. But this is not the last we'll hear of direct involvement from these asset holders.

A number of NGOs have started to push shareholder resolutions on environmental issues. The Interfaith Center on Corporate Responsibility, a coalition of 275 faith-based institutional investors, has cast a spotlight on oil and gas companies. Other groups, like the Boston-based coalition of investment and environmental groups called Ceres, have insisted on more transparency of the financial risks associated with climate change. Although these resolutions generally lose, they regularly get 20 to 40 percent support from shareholders.

To avoid continuing votes, a number of companies have given in. Six major oil and gas companies, including ChevronTexaco and Apache, agreed to acknowledge publicly the potential financial impacts of climate change, adopt plans to reduce greenhouse gas emissions, and change governance practices to give board-level support for these actions. Building on the success of these efforts, the Interfaith Center and others are expanding their targets to include smaller oil and gas companies and big players in financial services and real estate. Those service businesses will not be alone in facing questions once reserved for heavy industry. Beleaguered insurance giant AIG has recently responded to mounting shareholder pressure by creating a new climate strategy.

While the biggest splash is in the climate-change arena, activists are also using shareholder resolutions to push for action on a wide range of issues. Recently, companies as diverse as Avon, Dow, Wal-

MORE PRESSURE FOR DISCLOSURE

In 2002, a group of institutional investors, including ABN AMRO and Merrill Lynch, launched the Carbon Disclosure Project, which was simply a questionnaire sent to the world's 500 biggest companies. It asked them to document their emissions so investors could gauge climate change–related risk. Every year, over 60 percent of the companies post all their answers on the project's website (www.cdproject.net). The institutions joining forces behind this project represent over $20 trillion in assets.

Mart, and Whole Foods have faced resolutions about environmental health issues tied to certain chemicals and toxics. The Green Wave rolls on, not just among activists but also among investors deadly serious about the finances. Mindy Lubber, executive director of Ceres, notes: "This is not about progressive politics or conservative politics. It's not an activist campaign as much as it's a fiduciary duty to assess financial risk."

Insurers

At a recent speech, Peter Levene, chairman of insurance giant Lloyd's of London, spoke about the biggest risks facing the insurance industry. In a post-9/11 world, terrorism would seem to be number one, but here's what Levene said:

> Terrorism is a risk that is being taken care of in large part by governments right now. . . . The real issue for insurers is natural disasters, which are a very great concern. And the impact of those disasters has been increasing because the climate has been changing, which presents some very serious challenges for insurers.

Insurers don't mind risk, but they hate uncertainty. Their job is to predict the likelihood of something bad happening and spread the cost across all those facing the risk. To make money, they must accurately forecast the scale and frequency of the possible harm. Behind these companies lie the masters of risk management, reinsurers. An industry normally content to sit behind the scenes and provide insurance to the insurers, and thus take on a portion of *their* risk, reinsurers have gotten very vocal about environmental issues in general and about climate change in particular.

Big reinsurers like Munich Re and Swiss Re have good reason to worry. The total cost of natural disasters has risen rapidly in recent years. The 1990s saw more economic losses than the previous four decades combined, and the 2000s are not looking better. Super floods in Europe in 2002 cost $15 billion. The 2003 European heat wave killed 26,000 people and caused $16 billion in damage. In 2004 natural disasters cost the reinsurance industry $40 billion, not including the tsunami in South Asia. And in 2005, worldwide economic losses from natural disasters topped $200 billion.

Banks and Capital Markets

Over the last few years, banks have been waking up to the fact that the environmental and social risks on projects they lend money to, while hard to quantify, can be very damaging to their business. Default risk is an obvious problem, but risks to the bank's reputation are even more threatening. Again, we can thank some tenacious NGOs for making that connection clearer to international financiers. Organizations like Rainforest Action Network realized an obvious truth: Influence those who hold the purse strings, and you don't need to force change directly on the companies creating the problems. You don't like how a forestry company treats the forests and waterways near its mills? Then go after the people who give it the money to build those mills in the first place.

One of the first companies to respond to the early pressure and get ahead of the curve was ABN AMRO, a Dutch bank with over half a trillion euros in assets. ABN AMRO is roughly the twentieth largest bank in the world, and it's a true leader in trying to incorporate environmental thinking into the business. Executives at the bank initiated the environmental review obligations embedded in the Equator Principles and are pushing the envelope much further. As André Abadie, the head of the bank's Sustainable Business Advisory

THE EQUATOR PRINCIPLES

In 2003, ten global banks including Citigroup, Credit Suisse, and ABN AMRO, announced a new agreement they dubbed the "Equator Principles," creating new standards for how the banks make decisions about project finance loans. Big developers of massive projects like pipelines and power stations now need to prove that the environmental and social impacts of the construction have been thought through and will be mitigated. So intense has been the pressure to join the agreement that the signatories have already expanded from the original ten to over forty. These banks represent the vast majority of project financing worldwide. More and more projects are being rejected or modified. A pipeline in Peru, a mine in Romania, and other projects are being abandoned because environmental considerations were not taken seriously enough.

Group, told us, "the Equator Principles are just the tip of the iceberg."

Abadie's group was founded only a few years ago, but its importance in the company has grown very fast. His team is brought in to assess potential deals for environmental and social risks, only a fraction of which are project finance–related. Deal volume has doubled each of the last three years. Of the hundreds of deals reviewed during one recent year, the group restricted over 20 percent and nixed 15 percent outright. ABN AMRO is also building checks and warning flags into some of its automated risk assessment tools, so that assessment of environmental and social risk is ingrained deeply in the process. The bank has been publicly recognized for this deep and meaningful commitment, earning the World Environment Center's 2006 Gold Medal for International Corporate Achievement in Sustainable Development.

Banks are realizing that their lending portfolios are an enormous source of risk. Lending to an oil refinery, for example, may seem like a good idea today. But over the forty-year time frame of the investment, a lot about what the world values and expects could change, making a fossil fuel project a very bad investment. Even JPMorgan, a bit late to the party in 2005, told the world that it would "calculate in loan reviews the financial cost of greenhouse-gas emissions, such as the risk of a company losing business to a competitor with lower emissions because it has a better public standing."

Top Citigroup executives have also committed their bank to screen projects for adverse effects on natural habitats, refuse loans to companies with illegal logging violations, invest in renewable energy projects, and report on the greenhouse gas emissions from all power projects in the company's portfolio. Hedge funds, too, are getting into the act, wielding influence over governance and strategy deci-

THE GREENING OF FINANCIAL SERVICES

As the Green Wave rolls through the financial world, questions are being asked of borrowers of all sizes. Even small businesses must expect to be quizzed on their environmental impacts before receiving a loan.

sions at major corporations. We see this trend migrating rapidly through the world's economy, making the management of capital flows a leading edge of the Green Wave.

A TOOL TO MANAGE THIS COMPLEXITY: STAKEHOLDER MAPPING

As simple as it sounds, the best way to deal with all the different players out there is to take a structured approach. In Chapter 11, we discuss stakeholder strategy in more detail, but we start here with a simple tool for mapping 20 critical stakeholder groups (see figure).

Eco-Advantage Players

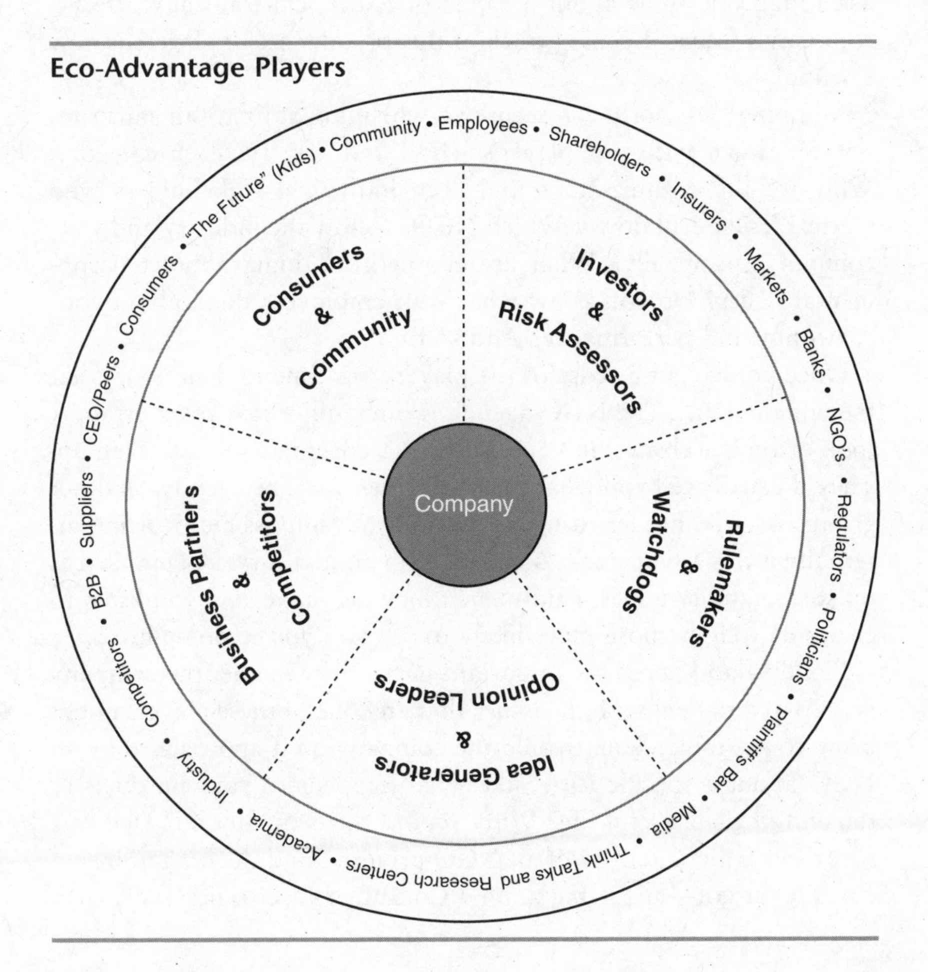

We use this tool to help companies develop forward-looking strategies for dealing with diverse and growing stakeholder pressures. But it also shows how much the world has changed. In traditional business analysis, the world of strategy involves a handful of key players that make up the value chain from suppliers to customers, plus owners (shareholders), rule-makers (regulators), and employees.

Today's world is almost unrecognizable from this traditional perspective. The "Five Forces" identified by Harvard Business School professor Michael Porter two decades ago still matter, and the players representing these forces are still very powerful. But the nature and ferocity of competition has changed, and business strategy has to be updated to keep pace. Environment-oriented stakeholders, who ask tough questions about a range of issues, can radically affect a company's future. Keeping track of this broader set of players is now essential.

Mapping this world can seem overwhelming at first. But start simply. Sit down with our player's wheel and ask, for each category: Who are the organizations and even individual stakeholders who matter? Then drill down: Which NGOs follow the industry and your company specifically? What are competitors doing about environmental issues? Do you know what your employees think about your environmental performance? And so on . . .

Once you've got a sense of the players, it's time to figure out what they might want. The NGO agenda is often quite easy to ascertain—look at their websites and at their media coverage. Or call them up. They'll usually tell you what their priorities are. The agendas of other groups may be harder to figure out, but not impossible. Starting internally is not a bad idea—Alcan does an annual survey of employees on sustainability issues. Painful as it may be, make sure you listen to critics as well as those more likely to tell you you're doing great.

Finally—and here's the important part—ask whether the company is prepared to deal with the issues that any one of these players might bring to the table. What would the company do if approached by an NGO about a specific issue, like what happens to your products at the end of their useful life? What *should* the company do? Dell was surprised at the intensity of an NGO protest—and the level of media coverage it got—at the big annual Consumer Electronics Show over the company's social and environmental practices. A year later when an NGO dumped a truckload of electronics in front of the annual shareholder meeting, the company was more prepared. Now, Dell maps stakeholders regularly and keeps a very close eye on external relations.

THE ECO-ADVANTAGE BOTTOM LINE

- The set of players affecting environmental strategy is growing in number, diversity, and power.
- Every company needs to watch the full stakeholder playing field carefully. Start by constructing a stakeholder map.
- Assess the influence and impact of the various players. But be careful: The ones you think are most influential may not be. Beware of underestimating others.
- Systematically review the players in each category as they relate to your company. Ask what they might want, and what the company's reaction could and should be.
- Look for opportunities to connect with critical players, even ones who might seem hostile.
- A "we'll cross that bridge when we come to it" approach invites trouble. Build relationships before you need them.

Part Two Strategies for Building Eco-Advantage

How do companies create an Eco-Advantage? To answer this question, we first had to ask a more basic one: How do companies create competitive advantage in general? In his seminal works on strategy, Harvard Business School's Michael Porter describes two basic categories of competitive advantage. A company can:

- Lower its costs compared with the competition.
- Differentiate its product on quality, features, or service.

Porter's work on competitiveness proved a useful starting point for analyzing the Eco-Advantage strategies we saw WaveRiders using.

Some costs are obvious and relatively short-term: inputs used, energy consumed, time and money spent on meeting regulatory requirements. More fundamentally, a great deal of pollution is waste and a function of outmoded production processes or poor product design. So improving the resource productivity of a business—the amount of material or energy

Strategy Framework

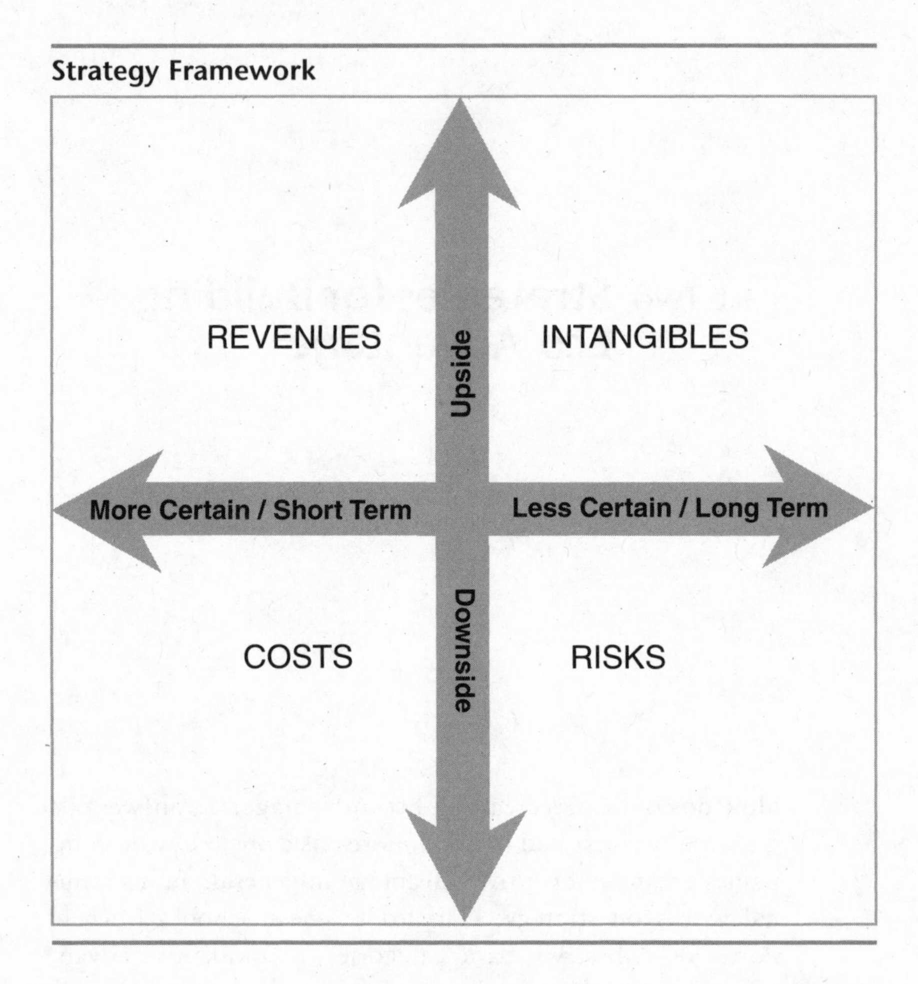

needed per unit of output—goes straight to the bottom line. Similarly, eliminating regulatory burdens by avoiding products, chemicals, or processes that require special care and documentation lowers overhead.

Companies that successfully manage environmental risks lower operating costs, reduce the cost of capital, drive up stock market valuations, and keep insurance premiums reasonable. They also avoid the indirect costs of business interruption and lost good will.

On the revenue side, the benefits of differentiation through good environmental stewardship are sometimes concrete—like commanding a price premium or just selling more—but are largely intangible: strengthened relationships with customers, employees, and other stakeholders. Some say that these intangibles are too vague to be measured, but they're wrong. How much does it cost to acquire a new customer to replace a lost one? That's the rough value of increasing loyalty. How about employee churn? If improving morale and employee engagement in the company's mission lowers turnover, how much would that save? And what about community support? What does it cost Intel in carrying costs, for example, if it can't build the next billion-dollar chip plant for twelve months because of community unease about how much water the company uses? These measurable gains make investments in intangible values more concrete.

To help us think through the environmental strategies companies use, or fail to use, we added one more dimension to the analysis. We asked ourselves whether a strategy was fairly certain or less certain to generate value. To oversimplify, we say that "certain" is roughly equivalent to the short-term and "less certain" to the long-term.

Take cost control versus risk management as an example. If you decrease waste in your system, you can be pretty sure how much you'll save. And you'll have an easier time selling the project internally. But what will it save the company to substitute a less toxic substance that costs more upfront? The risk is lower, but what is that worth? When does the benefit come? These questions are harder to answer, so risk control is less certain, although it often pays off more in the long run. The same holds true for the upside: It's easier (though not easy) to drive revenues than to increase brand value.

The Eco-Advantage Playbook

Through our interviews and research, we saw WaveRiders using eight fundamental strategies that accomplish one of the four overarching strategic tasks:

1. WaveRiders cut operational **costs** and reduce environmental expenses—like waste handling and regulatory burdens—throughout the value chain.
2. They identify and reduce environmental and regulatory **risks** in their operations, especially in their supply chains, to avoid costs and increase speed to market.
3. They find ways to drive **revenues** by designing and marketing

products that are environmentally superior and meet customer desires.

4. A few companies, most famously BP and GE, create **intangible brand value** by marketing their overall corporate greenness.

The full set of strategies, our Green-to-Gold Plays, defines the Eco-Advantage playbook. Smart companies use these strategies to convert environmental and sustainability thinking into profit.

In Chapter 5, we'll look at upside plays. But first, in Chapter 4, we focus on lowering both costs and risks, allowing for a much smoother ride on the Green Wave.

THE GREEN-TO-GOLD PLAYS

Managing the Downside (Chapter 4)

Cost

1. Eco-efficiency: Improve resource productivity.
2. Eco-expense reduction: Cut environmental costs and regulatory burden.
3. Value chain eco-efficiency: Lower costs upstream and downstream.

Risk

4. Eco-risk control: Manage environmentally driven business risk.

Building the Upside (Chapter 5)

Revenues

5. Eco-design: Meet customer environmental needs.
6. Eco-sales and marketing: Build product position and customer loyalty on green attributes.
7. Eco-defined new market space: Promote value innovation and develop breakthrough products.

Intangibles

8. Intangible value: Build corporate reputation and trusted brands.

Chapter 4 Managing the Downside

GREEN-TO-GOLD PLAY 1: ECO-EFFICIENCY—IMPROVE RESOURCE PRODUCTIVITY

Over the last fifteen years, chemical giant DuPont has cut its contribution to global warming by an astounding 72 percent. Half of the cuts came from changing only one process: the production of adipic acid. This modification eliminated emissions of nitrous oxide, a potent greenhouse gas that causes far more warming than carbon dioxide. The company also vowed to hold flat its energy use—the primary source of its greenhouse gas emissions—no matter how fast the company's top line grew. Through constant vigilance and innovation, the company found a hundred ways to get leaner and meet its energy targets. Over the past decade, this strategy has saved DuPont $2 *billion*.

That kind of dogged determination is typical of the smart firms we studied. WaveRiders get the same output with lower inputs. In improving resource productivity, their actions stand out as the classic win-win environmental strategy. Examples are plentiful:

- **Water:** Chipmaker AMD modified a "wet processing" tool to use fewer chemicals and, ironically, less water to clean silicon wafers. The process once used eighteen gallons of water per minute, and now it's fewer than six.
- **Material:** Timberland redesigned its shoe boxes to eliminate 15 percent of the material (which adds up when you ship over 25 million pairs per year).
- **Energy:** IBM recently overshot its five-year greenhouse gas reduction target, saving $115 million through energy-efficiency initiatives such as redesigning heating and cooling systems.

In our research, we've uncovered thousands of ways companies have reduced waste, saving both money and resources. Sometimes it's big initiatives like Dow Chemical's twenty-year-old Waste Reduction Always Pays (WRAP). Or it can be small changes like the computerized sprinkler system at the headquarters of software company Adobe Systems that checks the weather forecast before deciding to water the grass. Big or small, eco-efficiency has become a baseline element of smart business. But all movements begin somewhere. In this case, the idea of large-scale "pollution prevention" got its start in Minneapolis, Minnesota, with industrial giant 3M.

Pollution Prevention Pays

In 1975, Joe Ling, 3M's executive in charge of all things environmental, was busy complying with the relatively new laws of the land. His company was placing scrubbers on smokestacks to eliminate contaminants, treating effluents before releasing wastewater, and segregating solid waste so that some could be incinerated rather than just dumped. But wouldn't it be far easier, Joe thought, to eliminate the pollution before it happens? So he started a program that survives to this day, dubbed Pollution Prevention Pays (or 3P).

From the beginning, the program was unapologetic about one thing: Any idea that could reduce pollution should also save money. Executives today tell us that all 3P projects still live up to that ideal. "Anything not in a product is considered a cost. . . . it's a sign of poor quality," says Kathy Reed, 3M's top environmental executive. As 3M execs see it, *everything* coming out of a plant is either product, by-product (which can be reused or sold), or waste. Why, they ask, should there be any waste? And for thirty years, 3M management has been convinced that anything that increases its footprint—emissions, solid waste, energy or water use—is a sign of inefficiency.

3P is their answer. It's a program entrenched in the company's culture that encourages employees at all levels to rethink products and processes, no matter how small. Initially, Ling and his team were proud of the twenty waste-cutting, money-saving ideas their employees came up with. They saved many tons of pollutants and $11 million. 3P has since grown beyond even the most optimistic projections. Today the program claims a cumulative total of almost 5,000 projects and environmental savings of 2.2 *billion* pounds of pollutants. Emissions of volatile organic compounds alone have dropped from 70,000 tons in 1988 to fewer than 6,000 tons today.

The financial impact has been remarkable. 3M calculates that the company has achieved about $1 billion in first-year project savings. This is worth repeating—3M calculates only the *first year* of projected or actual savings from an eco-efficiency project. This overly conservative assumption keeps 3M honest and forces everyone in the company to look for ideas that have immediate benefits. And while understating the impact of the program, it shows how dramatic the gains from eco-efficiency can be.

After thirty years, 3M's 3P initiative is still generating new gains every year. 3M executives have set aggressive goals for the number of new 3P projects they'd like to see, but they haven't set monetary or even environmental goals. Just encourage people to look for new ideas and innovate, they think, and both the environmental benefits and the money will follow. Their experience bears this out.

Jim Omland runs five 3M plants that make medical tapes and industrial minerals. When he asked his employees to find three new 3P projects, he got some push-back. "They told me, 'But we've gotten all the savings we can here,' " Omland says, "and yet, when natural gas prices shot up and my business took a $10-million hit, suddenly my people found new ways to reduce natural gas use."

Over and over again, WaveRiders find that asking people to look at their work through an environmental lens leads to innovation

WHAT'S WRONG WITH ABATEMENT

Scrubbers on smokestacks are one symbol of a "we can fix it on the back end" attitude. But as 3M understands, scrubbers and similar technologies just shift problems from one place to another. The pollutants that scrubbers capture become sludge that still must be disposed of carefully or it creates water pollution. Or the sludge is incinerated, creating air pollution after all. As former 3M manager Thomas Zosel said, "All we're really doing is moving pollutants around in a circle."

3M's Pollution Prevention Pays program works so well because it asks people to stop problems before they start. Many WaveRiders told us that some of their biggest environmental slip-ups came from new abatement technologies that cost too much, didn't work as planned, or created more problems. Redesigning process and product to eliminate waste, rather than improving clean-up strategies, is as a central element of Eco-Advantage.

around waste reduction and resource productivity, which translates directly into Eco-Advantage.

Some companies are going beyond waste reduction and efficiency gains—and actually finding markets for their industrial by-products that would otherwise have been disposed of as waste. Rhone-Poulenc broke new ground in the 1990s when it found a market for the diacids that are a by-product of its nylon production. Today, many companies have adopted this spirit of "industrial ecology," where one firm's byproducts are another's inputs, and found ways to recapture and sell part of their waste stream.

Low-Hanging Fruit: Retrofits and Automation

In Chapters 7, 8, and 9, we will highlight a number of tools that companies use to find eco-efficiency. Life Cycle Assessments and Design for the Environment, for example, bring environmental thinking into practice. But sophisticated tools aren't the only way to foster eco-efficiencies. WaveRiders with a retail presence or large facilities often discover that the payback for installing new "green" lighting or other energy saving devices can have a high return on investment and pay back in months.

INFORMATION AGE ECO-EFFICIENCY

Opportunities for improved efficiency are much easier to find in today's digital world. Computers and information management systems make resource use and productivity easy to track and benchmark across facilities, products, and production lines. Comparative analysis of raw materials consumed, energy required, and waste generated simplifies the process of spotting best practices and capturing potential efficiency gains. E-mail and the Internet facilitate spreading these best practices across a company, speeding up feedback loops, and enhancing performance.

Digital technologies also create new eco-efficiency opportunities beyond the "factory gates." By bringing buyers and sellers together online, the Internet lowers search costs and makes markets possible that might never have existed. A growing number of waste exchange websites help companies "close loops" and find customers for their industrial byproducts.

At a micro-level, Dow Chemical set employees' computers to shut down when not in use. At the macro-level, Staples saved $6 million in two years with centralized controls for lighting, heat, and cooling at its 1,500 stores. And FedEx Kinko's retrofitted over 95 percent of its 1,000 branches with new energy-efficient overhead lighting and motion sensors to shut off lights when nobody was around. The company spent $3,000 to $10,000 per "center" and earned that back in energy savings in only twelve to eighteen months.

FedEx Kinko's Environmental Affairs Director, Larry Rogero, downplays his company's efficiency efforts. "Everyone does it, so it's not that innovative," he says. We disagree. Not every company looks for these simple energy-saving techniques, and very few retrofit 1,000 locations. FedEx Kinko's got down in the trenches and made real changes for a significant improvement in environmental impact and bottom line savings.

How Important Are These Cost Savings?

When we met with 3M executives, one thing was bothering us. 3M's net and operating margins are roughly the same as thirty years ago.

HARD TRADE-OFFS

Sometimes eco-efficiency comes at a price. Reducing waste along one environmental dimension can create problems elsewhere. A small Swiss manufacturer, Rohner Textil, designed a closed-loop water system to recycle this ever-more precious resource. Given water prices in Switzerland, the plan was projected to save a bundle. But the company soon realized that its new system significantly increased energy use, eliminating any savings, and CEO Albin Kaelin scrapped the program.

Environmental issues sometimes come into conflict with social considerations. Coca-Cola has faced significant public outcry in India over the discovery of trace pesticide residues in its products. After careful analysis, Coca-Cola discovered that the residues came from the sugar the company bought locally. One solution was to source sugar solely from outside India, but that would mean taking money and jobs away from Indian sugar cane farmers. In the end, Coca-Cola kept sourcing from Indian farmers and took on the cost of additional purification.

The moral: Before launching an eco-efficiency initiative, or acting quickly to reduce an environmental impact, look for unintended negative consequences.

So why don't we see the billion or so in eco-efficiency savings reflected in the company's margins over time? Their answer demonstrates just how important the 3P program has been to the company. 3M operates in many highly competitive businesses with eroding margins. Consistently finding ways to reduce costs has, in Kathy Reed's words, "kept us competitive and allowed us to stay in industrial businesses."

Others who have jumped on the environmental and cost efficiency bandwagon are even more direct. Ray Anderson, Interface's founder and chairman, told us that the company's $300 million in cost reductions from waste management and eco-efficiency saved the company. During the recession of the early 2000s, sales in Interface's primary market, the office flooring business, dropped by more than a third. "We wouldn't have made it" without those cost reductions, Anderson observed. Let's face it, efficiency is hard to make sexy, but higher margins make every CFO's and CEO's ears perk up. Business survival is pretty uplifting too.

GREEN-TO-GOLD PLAY 2: ECO-EXPENSE REDUCTION—CUT ENVIRONMENTAL COSTS AND REGULATORY BURDEN

The late 1980s were a wake-up call for DuPont. Public disclosure of environmental information was on the rise, particularly through the Toxics Release Inventory. The company discovered that it was one of the world's largest polluters, even though it was spending over $1 billion annually on waste treatment and pollution control. Management was shocked to discover how much money the chemicals going up the stack were costing them.

CEO Ed Woolard demanded that the company slash both emissions and costs. He set bold waste targets. "The goal is zero" became a DuPont mantra. When Woolard felt the company wasn't moving fast enough, he told executives, "If I have to shut a plant down to show how serious I am, I will."

They got the message. Today, DuPont's waste treatment and pollution control expenses are down to $400 million. Paul Tebo, former VP for Safety, Health, and Environment, estimates that those expenses otherwise would have grown to over $2 billion. That's a swing of $1.6 billion in annual costs for a company netting about $1.8 billion a year. Throw in the few hundred million per year on energy savings, and DuPont would roughly break even without its environmental efforts.

As with DuPont, so with every other business, particularly those that already spend millions on pollution control equipment. The scale can be daunting. Alcan's Dan Gagnier estimates that up to 20 percent of the $3 billion spent on a new aluminum smelter goes toward environmental equipment.

The first Green-to-Gold Play was about cutting costs by not wasting resources. This second one centers on the time and money consumed by pollution control and environmental management. In addition to millions of dollars or euros spent on waste disposal and pollution control equipment, we include the managerial time and money spent filling out forms, the sometimes crippling cost of fines for mismanaging environmental issues, and the general business slow-down caused by jumping regulatory hurdles.

Companies that tackle these expenses directly save hard cash. Fifteen years ago, furniture maker Herman Miller sent 41 million pounds of waste to landfills. Today, it's only 5 million pounds. Aggressive recycling and waste reduction efforts have saved the company over $1 million each year.

Anything a company can do to avoid regulations will lower operational costs and increase speed to market. Building a new facility, for example, requires countless permits. Using certain chemicals or going above an emissions threshold can trigger additional requirements. WaveRiders watch these levels closely and do everything they can to stay below the regulatory limits. If necessary, they redesign processes and products to get there. Seeing the business through the lens of environmental expenses can help companies find new, lower-cost, and faster ways of doing business.

GREEN-TO-GOLD PLAY 3: VALUE CHAIN ECO-EFFICIENCY—LOWER COSTS UPSTREAM AND DOWNSTREAM

Making shoes is a surprisingly toxic business. Aside from the materials themselves, the adhesives that connect them are made from chemicals that are known dangers to the cardiac, respiratory, and nervous systems. One pair of running shoes isn't going to land you in the hospital, but workers in the industry face real risks.

Convinced that it needed to rethink the traditional industry reliance on toxic chemicals, Timberland became the first footwear company to test new water-based adhesives on non-athletic shoes (Nike and others had already made some strides in the "white shoe," or athletic, side of the industry). Making the change required the company to work closely with Asian suppliers.

Common sense would seem to suggest that Timberland's detoxification initiative would cost its suppliers a bundle. And during the test phase it has been more expensive—the new adhesives cost more since economies of scale haven't kicked in yet. But over time, Timberland fully expects the process to be at least cost neutral for its business and a money saver for the full value chain. Here's why: Water-based adhesives eliminate almost entirely the supplier's expense for handling hazardous materials, including waste disposal, insurance, and training. Manufacturing expenses already dropped during testing since the water-based adhesives go on with one coat instead of two, and the application equipment requires much less cleaning. The suppliers can run longer without interruption. The change saved both labor cost and time. But will Timberland be able to capture these supplier savings down the road? That's the challenge of Green-to-Gold Play 3.

Clearly, it's not easy. If Timberland finds it difficult to ask suppliers to share the savings, imagine asking customers to do the same. Take an example from DuPont's innovative work on automotive paints. The company's SuperSolids technology, which DaimlerChrysler is currently testing, reduces some hazardous emissions from the coating process by up to 80 percent. DuPont estimates that the technology can save the auto companies $20 million per plant in emissions control equipment and operational expense. So can DuPont expect a piece of these savings? Perhaps not, but over time, this strategy helps DuPont win market share, driving new revenues.

Many companies have found ways to lower value chain costs by cutting the environmental and financial expenses of product distribution. Anyone who has bought an assemble-it-yourself product from IKEA knows how much stuff IKEA fits into a box. The company is justifiably proud of what it calls its "flat packaging." These efforts to squeeze millimeters out of every box have allowed IKEA to pack its trucks and trains much tighter. In some cases, the company has achieved a 50 percent increase in fill rate. That kind of smart packing saves up to 15 percent on fuel per item—a striking Eco-Advantage—and it inspires workers to stretch the envelope even

FILLING THE TRUCKS

It's a shockingly simple value chain efficiency play: Fill your trucks as full as possible. For example, Dell has upped its average truck load from 18,000 to 22,000 pounds and worked with UPS to optimize delivery strategies. And one of 3M's recent 3P award winners was an innovative system developed by a French employee to install adjustable decks in trucks. Placing pallets on two levels allowed just one 3M facility to reduce the number of daily truckloads by 40 percent and save $110,000 per year.

more. IKEA employee Erik Andersson noticed that the company's 88-centimeter KLIPPAN sofa was being shipped in a 91-centimeter box. Redesigning the packaging to cut out just one of those centimeters allowed IKEA to fit four more sofas on each trailer.

GREEN-TO-GOLD PLAY 4: ECO-RISK CONTROL—MANAGE ENVIRONMENTALLY DRIVEN BUSINESS RISK

For many years, kids have been happy to find prizes or toys in cereal boxes. In the summer of 2004, a seemingly routine cross-promotion with the new blockbuster movie *Spider-Man 2* turned ugly fast for Kellogg Company, maker of Rice Krispies and Pop-Tarts. As the Associated Press put it cheekily, Kellogg was "caught in a web of criticism" over the new Spidey Signals toy found in boxes all over the country. Apparently each electronic toy had a surprise of its own: the small battery that gave it power contained toxic mercury.

Such "button" batteries are quite common, but a few states have banned mercury-powered toys. Kellogg suddenly found itself under public attack for putting mercury near kids' food. After hearing from state attorneys general from New York, Connecticut, and New Hampshire, the company offered to send a prepaid return envelope to each of the 17 *million* customers who had received the toy. And Kellogg committed to never again using the offending batteries.

Meanwhile, at the Illinois headquarters of another food giant that often plays with toys, executives were breathing a huge sigh of relief. McDonald's had dodged this same bullet, but not by accident. A few years before Kellogg's PR problems, McDonald's had identified button batteries as a growing risk and eliminated mercury entirely from all Happy Meal toys.

We interviewed McDonald's executives just days after the Kellogg incident. The company's executives, we discovered, have developed a strategic approach to identifying and reducing risks to the brand, by far the company's most valuable asset. As guardians of a megabrand with intangible value in the tens of billions of dollars, they work hard to reduce surprises and enterprise risk, including environmental risk. After years of facing pressure over everything from litter and packaging to mad cow disease, McDonald's decided get ahead of the curve.

Through a process of "anticipatory issues management," the company studies environmental and social trends to identify potential dangers to its business. Early on, one of the emerging threats they focused on was the then-obscure issue of mercury in batteries. McDonald's calculated the downside to be high and the corrective cost low, so a few years before Kellogg had its Spidey surprise, McDonald's used its market clout to pressure suppliers to find different options. By the time states started to regulate mercury in toys that come near food, McDonald's was long gone. The batteries never became a costly, brand-damaging problem because the company systematically identified the risk and then avoided it.

Problems that *don't* arise are a strange kind of success. Bob Langert, McDonald's Director for Social Responsibility, told us, "I'm proud of what we've done, but nobody knows about it." In this case, we're sure that's exactly how shareholders would want it.

BUILDING A TRUST BANK

For Bob Langert and the McDonald's team, risk management is about more than controlling the downside. "Sure, the risk comes if you're not doing the right thing," Langert said, "but the upside is building a 'trust bank' with customers. It's very hard to gain, but easy to lose. . . . The more you build it, the more you build loyalty to the brand. I'm convinced there's a real opportunity here."

Finding the Risk Before It Finds You

Oprah Winfrey has serious market power. She can pluck a book from obscurity and make it a best seller. On her recommendation, millions of people change their buying patterns. In 1996, Oprah interviewed a vegetarian activist on her show and declared that what she heard "just stopped me cold from eating another burger." Beef prices dropped over 10 percent the next day.

Business risk comes in the strangest forms. For beef producers and burger sellers like McDonald's, the Oprah incident must have seemed like a freak accident—a meteor shower from out of the blue. But was it unpredictable? Not really. The vectors were all in place. Oprah just put them together for her viewing public.

Smart companies use many methods to identify risks, even hard-to-spot ones. Shell uses scenario planning to paint pictures of possible futures. IKEA does exhaustive supply chain auditing. McDonald's draws top managers in to regular risk reviews.

Looking for environmental risks requires going far beyond the traditional company boundaries. Risks may arise upstream (with suppliers) or downstream (with customers).

Imagine that a big-box retailer has an Asian supplier of leather coats that's dumping hazardous waste from its tanning operation into a local river. An enterprising NGO has taken pictures of this illegal dumping and posted them on the web. The story begins to draw attention in the U.S. and European press. Customers won't remember the name of the small tannery, just the big-brand retailer which will take the hit for being an environmental bad guy. Scrambling to address this problem after the story has broken leaves the company in a Humpty-Dumpty situation: A reputation, once shattered, can't easily be put back together.

WHAT IS BUSINESS RISK?

It's easy to say that every company should reduce environmental risk. But what exactly should managers be looking for? We think of business or enterprise risk as the chance that something will change "business as usual" into something quite different. The experts at the Institute of Risk Management in the United Kingdom lay out four broad categories:

- **Financial:** interest and exchange rates, liquidity and cash flow
- **Strategic:** competitors, industry dynamics, customer changes
- **Operational:** supply chain, regulatory
- **Hazard:** a wild card of natural events, environment, employees, and so on.

Environmental risk plugs into all these categories: Liabilities for spills or other incidents affect the financial prospects, customer needs shift the strategic landscape rapidly (better gas mileage in cars for example), and the specter of tighter regulations or supply chain problems is an operational risk.

So WaveRiders find issues before the problems find them. And they examine not only the supply chain, but the entire value chain. We suggest a few big-picture questions that can help companies get a handle on environmental impacts (see table).

At the nuts-and-bolts level, identifying enterprise risk means understanding exactly how a company affects the environment and how the constraints of nature affect the company. Our AUDIO scan (from Chapter 2) can help a company identify where environmental issues touch the business along the value chain. In Chapter 7, we'll discuss

Identifying Environmental Risk

Value Chain Phase	Sample Questions to Help Identify Environmental Risk
Company Operations	— How big is our environmental footprint? — What resources are we most dependent on (energy, water, materials), and how much do we use? — What emissions do we release into the air or water? — How do we dispose of waste? — How up-to-date is our environmental management system? — What are our chances of a spill, leak, or release of hazardous materials? — Have others in our industry had problems? — What local, state, federal, or international regulations apply to our business? Are we in full compliance? Are these requirements getting tighter?
Upstream	— What resources are our suppliers most dependent on? Are they abundant or constrained, now and in the near future? — Do our suppliers pollute? Do they meet all applicable laws? Will legal requirements get tighter for them? — What substances go into the products suppliers sell to us? Are they toxic?
Downstream	— How much energy (or water or other resources) does our product require customers to use? — Are there hazardous substances in our products? — What do customers do with our products when they are done with them? What would happen if we were required to take the products back?

other tools that help companies spot risks, even those that may be over the horizon.

Thinking Ahead: Go Beyond Compliance for Competitive Advantage

By the late 1990s, the McDonald's team in Hungary could see the future of recycling regulation. Western European countries already had highly developed waste-handling systems. With Hungary preparing to join the European Union, more stringent rules were on the way. Rather than wait for a government mandate, McDonald's Hungary management decided to build a country-wide, custom waste-handling system.

When the national recycling legislation passed, companies were "asked" to join and foot the bill for the new system. High fees and low initial service levels weren't much of a draw, but most companies had no real choice. McDonald's, however, had options. "McDonald's Hungary used its own system," EU Environmental Manager Else Krueck told us. "It was less expensive than the national one and tailored to the restaurants' waste stream." So the home-grown system was better and cheaper.

WaveRiders realize that getting ahead of regulations can save money and time, as well as reduce hassles. Privately held SC Johnson has quietly reformulated market-leading products such as Windex, Drano, Pledge, and Ziploc to reduce the use of some chemicals, in particular eliminating, "persistent, bioaccumulative, and toxic" substances sometimes called PBTs.

All SC Johnson products are subjected to an internal process called Greenlist that gives every ingredient a score based on environmental attributes such as toxicity and biodegradability. SC Johnson has evaluated over 3,000 raw materials—far more than the federal government has rated under its toxics laws. Similarly, Nokia has reviewed 30,000 components and removed some materials from its products.

Dave Long, the SC Johnson executive who manages the Greenlist program, says the upshot is that the company is affected much less by new regulations than its competitors. "There's lots of scrambling within the industry when new regulations come down," he says. "When a new detergent regulation was passed in the European

GETTING READY FOR THE BIG ONE: CLIMATE CHANGE REGULATIONS

With climate change regulations emerging all over the world, smart companies are preparing for this future now, even in places such as the United States where mandates are not yet in place.

Getting a clear read on emissions at the facility level is a good place to start. Chipmaker AMD released its first Global Climate Protection Plan in 2001. This annual report describes the company's emissions by site and provides examples of projects in the works to reduce the total. Such a corporate climate strategy might have seemed odd a few years back, but today it's becoming both normal and expected.

As we mentioned in Chapter 3, the Carbon Disclosure Project, backed by some of the world's largest institutional investors, is asking companies a simple question: "What's your climate change plan?" What they really mean is, "What are you going to do if and when tighter regulations come down the pike?" Saying "I don't know" is no longer an option.

"We are in a carbon-constrained world *now*," GE's CEO Jeff Immelt has observed, "Tomorrow is today."

Union, because of Greenlist we had already reformulated our products to use surfactants that complied with the law. So the impact of the law was minimal."

Remember REACH, the onerous EU regulation that the chemical industry says will cost billions and destroy companies? Long is unfazed. "We're in good shape with REACH because several of the criteria are around toxics that we've already eliminated from our raw materials." Regulations are not so scary when they don't apply to you anymore.

Indeed, stronger regulations are welcomed by companies that are already beyond compliance. Stricter laws impose costs on the less-prepared competitors and potentially could keep them out of a market space for years on end.

In 1999, Swedish appliance manufacturer Electrolux announced a partnership with Toshiba to "develop energy-saving technology to prepare for the expected introduction of stricter global environmental regulations." That's looking ahead, but even smaller-scale local and national regulations can change marketplaces. Japan's "Top Runner" product labeling program shows customers the "total cost" of an appliance—the list price plus ten years of electricity to power the device. Since Electrolux makes some of the world's most efficient appliances, the company is clearly well-positioned for a world where Japan's eco-labeling regulations are the norm.

BP also started getting ready early. BP discovered $1.5 billion in efficiency savings by internally trading greenhouse gas emissions between business units (more on this later). The company's experience helped it shape the United Kingdom's emissions trading system, and then the European Union's. As BP's CEO Lord John Browne said, getting ahead of the curve means the company "gains a seat at the table and a chance to influence future rules."

Similarly, Nokia found it very useful to prepare its business for coming laws like the regulation on hazardous substances and, in particular, take-back laws that make manufacturers deal with their products when customers are done with them. Moving quickly, well ahead of regulations, allowed Nokia to pilot ideas and work out the kinks in its system. Like BP, it also gave the company a role with authorities in helping to shape the coming laws.

SEEKING ENVIRONMENT-BASED COMPETITIVE ADVANTAGE IS OK

Not infrequently we hear executives worrying that competing on environmental factors will be considered unseemly. We understand the sentiment, but executives needn't be shy about profiting from doing the right thing. If an electronics producer finds a way to make its products without heavy metals, why share that with the competition? Why not use the Eco-Advantage to stick it to competitors? If there are stakeholders that object, we're not sure who they are. NGOs are happy to see the cleaner companies gain the upper hand over the dirtier ones, which moves an industry toward greener solutions as surely as protests and regulations. And employees and shareholders certainly won't mind if environmental strategy translates into higher profits.

Advanced Strategy: Lobby for Stricter Regulations for Competitive Advantage

The fact that companies can influence the course of government policy is widely understood. Many companies invest vast sums in lobbyists, industry associations, and campaign contributions, all to wield influence over the political process. What is curious is that almost all lobbying efforts are aimed at *stopping* new regulations. Yet new regulations create winners as well as losers. Those best positioned to respond to new rules will be relatively advantaged by a changed playing field.

Far more often than they currently do, companies should ask for *stricter* regulations. Sure, it can be risky, but in the right circumstances, it's a powerful play that can yield significant advantage. Champion Paper thrived, for example, when its competitors faced new restrictions on timber cutting due to concern about endangered spotted owls in northwestern forests. DuPont gained market share (and profits) when the Montreal Protocol phased out production of ozone-depleting chlorofluorocarbons (CFCs). With $500 million in CFC-based revenues, DuPont initially fought the phase-out until it realized that it would make even more money in the CFC-substitute market.

THE ECO-ADVANTAGE BOTTOM LINE

Look to reduce costs by:

- Eliminating waste and promoting eco-efficiency
- Cutting disposal costs and regulatory compliance expenses
- Capturing the value of reduced environmental burdens up and down the value chain

Control environmental risk by:

- Anticipating environmental issues and addressing them
- Staying ahead of new regulatory requirements
- Managing government mandates to gain a relative advantage in the marketplace

Chapter 5 Building the Upside

"When I was made CEO, I never imagined I'd be talking about the environment," GE's Jeff Immelt said during the launch of the company's ecomagination initiative. Talk about an understatement! The previous CEO, Jack Welch, had a fiery relationship with regulators and NGOs. Welch battled the government for years over GE's responsibility for toxic PCBs found in the Hudson and Housatonic rivers. Yet just a few years later, his hand-picked successor declared that environmental goods and services would be a centerpiece of GE's business strategy.

The jury is still out on the long-run effectiveness of GE's ecomagination campaign, but the thinking and strategy behind it perfectly demonstrate the upside set of Green-to-Gold Plays. Ecomagination is a multipronged initiative, part image advertising, part straight-up product marketing, and part product innovation. At the core—and Immelt has made this very clear—it's about top-line growth. Early results are very promising: GE booked a $4 billion increase in sales of environmental products in the first year of the program.

Environmental strategy has been on a long march for the past forty years, from a tactical focus on compliance, to an additional—but still tactical—emphasis on costs and efficiency, to a more strategic view centered on growth opportunities. More and more companies now see the top-line potential from artfully managing the pressures of the Green Wave.

The four Green-to-Gold Plays set forth in this chapter are about growth—of sales, of brand value, and of stakeholders' trust. The strategies in this chapter focus on developing new products based on meeting customer needs, marketing the environmental aspects of those products, creating a new market space (or "value innovation"), and building corporate image around a company's commitment to being green.

GREEN-TO-GOLD PLAY 5: ECO-DESIGN— MEET CUSTOMER ENVIRONMENTAL NEEDS

Remember electric cars? Or the first wave of energy-saving light bulbs? These green products were brought to market by smart, successful companies like GM, Ford, Philips, and GE to satisfy environmentally driven customers. Who could blame them? For thirty years, surveys have shown that customers care about environmental issues. Yet despite what customers say in theory, when faced with an actual product with a higher price tag, they often don't buy (as all the companies above quickly discovered).

Some of these products failed because they didn't really meet a customer need at the right price, others failed because of ineffective positioning or marketing. Identifying customer needs or desires and designing a product to meet them is never easy. With growing environmental consciousness, the opportunities to seize Eco-Advantage through green marketing are expanding.

Lowering Their Burden: Make Your Customers' Environmental Problems Your Own

What exactly does it mean to say a product or service was eco-designed? The short answer is that the item was developed in a way that reduces environmental impacts for someone somewhere in its life-cycle journey from supplier inputs to product to end-of-life disposal. Often "Design for the Environment" helps customers lower *their* footprint and related costs—benefits that can justify price premiums, drive increased market share, and strengthen customer loyalty. At the heart of this Green-to-Gold Play are efforts to lower energy use, eliminate waste, or reduce product toxicity.

There are countless ways to help customers improve their eco-efficiency. Because creativity is the key to finding ways to cut waste or improve resource productivity, small and nimble businesses with an entrepreneurial spirit can often profit. IdleAire Technologies, for example, developed a service at truck stops that pumps electric power, heat, air conditioning, cable television, and high-speed Internet into the parked vehicles. This external supply allows the truckers to shut off their engines rather than keeping them idling all night. The service saves fuel, reduces engine wear and tear, and costs the truckers much less than the fuel needed to idle. And if implemented broadly, IdleAire's efficiency innovation could eliminate 34 million tons of greenhouse gases per year.

Small businesses are also helping the largest buyers of all, governments, to reduce their environmental impacts. Tiny Seahorse Power Company in Massachusetts makes a new kind of trash can called BigBelly. This solar-powered hi-tech container automatically compresses the garbage, reducing the number of trips trucks have to make for pick up. Customers like New York City and the U.S. Forest Service can send out fewer trucks and burn less gas.

Large technology companies are getting into the act as well. With much fanfare, Sun Microsystems launched a "green server" on a chip which reduces power consumption and cooling requirements. As CEO Scott McNealy said in his pitch to the computing world, "Sustainable growth strategies can help companies dramatically cut costs. . . . Sun is addressing energy and resource efficiency, power consumption and waste management, as we help businesses and our employees meet the challenges created by the evolving role of technology in our everyday lives."

An equally big opportunity waits at the back end of the technology life cycle: helping customers deal with product disposal. Dell's Asset Recovery System offers a valuable example of this play in action. Since computers turn over quickly, companies face real challenges over environmental and data liabilities as they look to dispose of obsolete equipment. Dell has stepped in to help customers deal with both the software and environmental clean-up they need. And it's making money doing it.

For roughly $25 per piece of equipment, Dell will come to your office and take used computers away. Dell first performs a "destructive data overwrite" to eliminate all digital information on the computer, then dismantles the machine. Dell refurbishes and reuses some parts, and recycles the plastics. In the end, just one percent of the old computer's volume goes to the landfill.

This multifaceted service improves the customer relationship and can drive sales. As Dell discovered, the take-back role comes, conveniently, when Dell is delivering the next generation of equipment. Company execs would be happy if this service were merely breaking even, but Dell is making money on it. They seem a bit sheepish about doing so. We see no need for apologies.

Three Lessons Learned on Driving Revenues with Eco-Design

Eco-design can be tricky. Companies have failed more than they've succeeded with this Green-to-Gold Play. Our research shows that companies can avoid the worst stumbles by following a few simple lessons.

MEET A NEED THAT ACTUALLY EXISTS

In the 1990s, DuPont's engineers were trying to "close the loop" in their polyester businesses. They invented a new technique for recycling polyester, dubbed Petra Tech, which "unzipped" the molecule and created new polyester from old materials. In theory, taking old product off customers' hands added value and lowered their costs. But unlike the challenges carpet companies or printers faced with toxic dyes and solvents, polyester disposal was not a big problem for customers. The recycled polyester actually cost more than virgin polyester, which was fairly cheap. In short, there was no compelling customer value proposition.

An innovative, environmentally sound process or product can be exciting for an organization. But what if the problem it solves is not one that any customer has? The lessons: Don't get caught up in the technology and forget to make the business case. And don't suppose that what is good for your company will necessarily be valuable to the customer.

DON'T IGNORE OTHER NEEDS OF THE CUSTOMER

It's very easy to get over-excited about a hot solution to an environmental problem and forget that the product still needs to do what it's supposed to do. Early in their quest to eliminate the use of solvents (and the nasty volatile organic compounds they produce), 3M scientists found a way to make magnetic audio tapes using water-based coatings. Unfortunately, the newly formulated product, which they could pitch as VOC-free, had a serious problem. The temperature range for the new tape was not as wide as the traditional product. In fact, the VOC-free tapes often melted under normal use. Oops.

Sometimes a product's functionality lies in the service it provides. McDonald's experimented with serving coffee in reusable mugs instead of disposable cups. But customers wanted to walk out the door with their coffee, not wait around to drink it so they could return the mug. They were paying for mobility as much as coffee.

PAY ATTENTION TO YOUR OWN COSTS

Even if a company identifies a need, filling it may be too expensive. When a nurse asked 3M why the packaging for one of its medical products was not recyclable, product managers took the request seriously. But changing packaging for a medical product, they realized, is a big deal, requiring lots of testing with many regulatory hurdles. The cost was just too great for the potentially modest environmental benefit.

The oil giants are facing this dilemma with new fuels. The extra refining to produce cleaner-burning fuels costs more and increases refinery emissions. While satisfying a customer's environmental needs can be very valuable, companies need to look first for unexpected costs and unintended consequences.

GREEN-TO-GOLD PLAY 6: ECO-SALES AND MARKETING—BUILD PRODUCT POSITION AND CUSTOMER LOYALTY ON GREEN ATTRIBUTES

Not every customer wants eco-friendly goods. But some do. And every day more consumers are including environmental factors in their buying equation. In parallel, companies are finding that there is money to be made from meeting the growing demand for green products. We've seen many examples at the leading edge of this trend:

- Melitta sells brown (unbleached) coffee filters alongside its traditional white ones. A certain percent of the coffee-brewing public wants to avoid the trace chemicals that might otherwise leach into their morning cup of joe.
- Whole Foods and other chains that focus on organic food are expanding rapidly, and Stop & Shop's Nature's Promise line of organic products is booming. Many such products now sell at substantial premiums. For example, organic milk often fetches more than double the price of regular milk and demand continues to grow.
- After some lean years, The Body Shop has begun to profit from the fact that green is increasingly cool. Other personal care companies have learned from The Body Shop's ups and downs and many are capitalizing on the growth of this market niche. Bath and Body Works' eco-friendly Pure Simplicity product line, for example, has seen a surge in demand.

When and Where Green Marketing Works

For a primer on green marketing, it's hard to top Shell Oil's experience marketing a new, cleaner-burning gasoline in two very different countries. Mark Weintraub, Shell's Director of Sustainable Development Strategy, told us that the company used a "sustainable development lens" to identify a need for cleaner fuels in Thailand. As in much of the rest of Asia, the combination of dense cities and high traffic volume was wreaking havoc on air quality in Bangkok and elsewhere. A fuel that would burn cleaner, producing less sulfur and other harmful emissions, might serve a real need.

In an example of good eco-design, Shell developed just such a fuel by converting natural gas to a zero-sulfur liquid and then mixing it with regular diesel. Today, the company sells the blend in Thailand under the brand name Pura. It's marketed as a way to reduce pollution and help engines run cleaner and last longer. Even though Shell charges a premium, Pura has grabbed market share and sales have been strong. The launch, in short, was a complete success.

Logically, Shell thought it could roll Pura out with the same pitch in other regions, but the launch back home in the Netherlands fell flat. Why? Shell later realized that stressing how cleaner burning fuel protects a car's engine was not resonating in Holland. That was far more important in Thailand where people worry more about gasoline quality and the effect of impurities on engine performance and life.

The green pitch never really resonated in Holland either even though the country is chock full of consumers who say they'll buy green. The need to clean the local city air is just less pressing than it is in Asia. In the end, Shell re-launched Pura at home under the name V-Power and marketed it by stressing enhanced engine power.

Shell's experience is not unusual. The green pitch is complicated. In just a few markets, the public instantly understands the environmental benefits and pays more for it. When that happens, green really is gold. Witness the incredible growth of the organic foods market. One WaveRider, Clif Bar, has latched on to this trend, moving all its core products, such as energy bars, toward organic ingredients.

ANOTHER DOOR TO SALES

In a world of tougher competition, any additional connection to customers can help to cement relationships. GE once asked 3M to share some of its thinking on "green chemistry" and the unique environmental challenges some products present. By sharing its world-class environmental thinking, 3M increased a connection with a major customer. As 3M's Kathy Reed told us, "Our environmental, health, and safety knowledge is another door to sales." Similarly, customers of Latin American conglomerate GrupoNueva often ask the company for help with their own environmental practices. In sharing this information, GrupoNueva solidifies its role as a business partner.

Three Lessons Learned on Selling Environmental Virtue

GREEN ATTRIBUTES CANNOT STAND ALONE

Selling a product on its environmental qualities alone is a recipe for trouble. If you have a new product that's cleaner and greener, marketing these advantages can make sense. But be careful. Customers need other reasons to buy. Price, quality, and service will remain core concerns for most of them.

A small niche in any market will want to hear the green pitch. But, as Shell's Mark Weintraub puts it, "A lot more consumers . . . are interested in the green attribute if it's the second or third 'button' you push—tell them this is higher quality product that will protect your engine and oh, by the way, it's better for the environment. That 'by the way' helps."

THE THIRD BUTTON

Marketing the green aspects of a product can be a tough proposition. Most successful green marketing starts with the traditional selling points—price, quality, or performance—and only then mentions environmental attributes. Almost always, green should not be the first button to push.

One way to signal that a product has environmental advantages without turning to green marketing is through certification and eco-labels (see figure). In a number of countries, labels that certify the green credentials can do the talking for you. Scandinavia permits environmentally superior products to be marked with the Nordic Swan logo, and Germany has its Blue Angel. In the United States, organic foods benefit from the USDA organic stamp. Antron, a maker of carpet fibers, saw a $4-million bump in sales when it became the first product in the commercial interiors industry to be certified by Scientific Certification Systems as an "Environmentally Preferable Product."

In some industries, expensive certifications can become the ante to play the game. A company like Chiquita really had no choice. Partnering with the Rainforest Alliance and radically changing the way

Eco-Labeling around the World

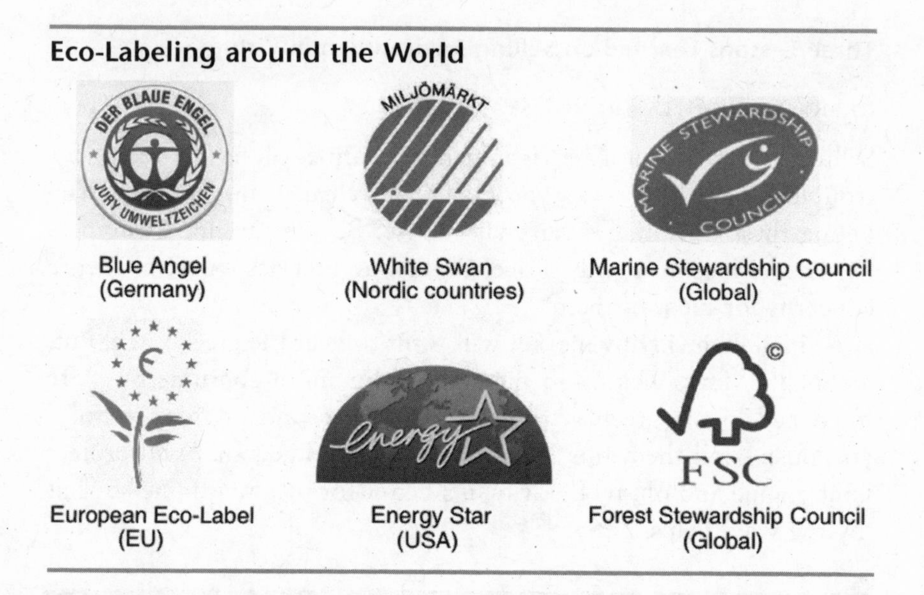

it grows bananas became essential for the company to serve the needs of customers, particularly in Europe. And increasingly, American electronics and appliance buyers look for the U.S. government's Energy Star label as a quick way to be sure they are getting an energy-efficient product.

As consumers demand more information on the products they buy, many companies have set up websites with facts, figures, and analysis concerning the environmental attributes of their goods. Others are starting to provide detailed eco-labels. Timberland, for example, is rolling out a new design element on its shoe boxes: a table that looks like the nutritional content label you find on food. The label tells customers, among other things, how much energy was used to produce the shoe.

Who does the certification and labeling, and on what basis, can be contentious. In some cases, governments set the standards. Other eco-claims are self-awarded. Still others are established by private entities such as the sustainable fish label issued by the Marine Stewardship Council or the sustainable wood certification offered by the Forest Stewardship Council. To achieve the environmental advances required for certification, companies often find collaboration with a third party helpful. Chiquita's work with Rainforest Alliance on banana farms shows how these partnerships can work.

GREEN PROTECTIONISM

Eco-labels can provide legitimate environmental information to a demanding public. But they also can be used as a trade barrier, disadvantaging competitors in the marketplace. For example, in the European beef market, local producers have tried to seize market share by seeking to have U.S. beef imports labeled as "hormone-treated."

Green protectionism can take other forms as well. Ontario at one time required that all beer be sold in returnable glass bottles. It sounds eco-friendly, but this recycling mandate offered a market advantage to Molsen and other Canadian brewers who used glass bottles as a matter of course. U.S. beer companies, who mostly sold their product in easier-to-recycle aluminum cans, got the short end of the deal.

The bottom line: Watch out for trade barriers and market-entry obstacles in the guise of environmental standards.

TALK TO DIFFERENT NICHES DIFFERENTLY

Market segmentation is nothing new. But with environmental issues, the differences in attitudes can be profound. Monsanto ran smack into a wall when it tried to bring biotechnology to Europe. United States customers didn't seem to flinch at the idea of food based on genetically modified organisms, but EU customers reacted so badly it nearly sank the company.

To get green consumers to buy, you've got to speak their language. WaveRiders recognize the need for tailored pitches. Office Depot developed a catalog dedicated to green products, including all the recycled paper and remanufactured toner cartridges a budding office environmentalist could ask for.

In the business-to-business market, the key is not just to talk to the right customers, but also to talk in the right way. A sales force that's not trained on why the environmental product is better can stop any launch in its tracks. Greener products often cost more up front, for instance, but end up saving the customer money down the road. Sales people need to understand this positioning. When we asked the head of sales at one of our WaveRiders if customers "got it," he laughed, saying, "You could ask if *my* sales guys even get it."

Sometimes, less is more. While Interface Flooring was going through its decade-long transformation to a sustainability focused company, chairman Ray Anderson worried about pitching the change before the company was clear on the message. "We forbade the sales force from talking about our green efforts for nine years," Anderson told us. "It's the kiss of death when words get ahead of deeds because customers see through this."

DON'T EXPECT A PRICE PREMIUM

Corporate strategy 101 tells us that a company can drive revenues by increasing price or volume. With green products, volume is a much safer route. Price premiums are rare. And they are likely possible only for truly innovative products that redefine the market space in some fundamental way—like the Toyota Prius, which we'll get to in a moment.

In any market, some customers will pay more for the green option. Shell's Weintraub suggests that this segment is about 5 percent of customers, while another WaveRider executive more pessimistically pegged the figure closer to 1 percent. Polls suggest that the figure can go as high as 10 to 20 percent in some markets, but don't bank on a big price premium unless you've got a very special product.

What do all these lessons on pitching green products have in common? They basically say the same thing: *You can't ignore the core business issues that accompany any product development and launch.* For eco-design and green marketing, as for other business initiatives, success stems from expertly handling all the regular blocking and tackling—identify a customer need, keep costs down, and meet performance and price expectations.

Every company offering a green product is also fighting a legacy issue: Some customers think "green" means poor quality or less functionality. This concern is not baseless. Electric cars didn't go very far or very fast, and compact fluorescent light bulbs created a harsh white light. In both cases, newer versions of the product have solved the problems. But the damage has already been done.

Even products that are environmentally leaps and bounds beyond what's out there need to get the basics right. And they must have other selling points as well.

GREEN-TO-GOLD PLAY 7: ECO-DEFINED NEW MARKET SPACE—PROMOTE VALUE INNOVATION AND DEVELOP BREAKTHROUGH PRODUCTS

In 1993, Toyota set out to design the "21st century car." During internal brainstorming sessions on what the next century held, engineers hit on two phrases: "natural resources" and "environment." They made environmental performance the focus of the new car, instead of traditional selling points like size or speed. Over the next decade, as oil prices rose fast, the logic of an energy-efficient car seemed obvious. But at the time, Toyota's strategy was very risky and the targets seemingly impossible.

First, top management set a goal that the new car would double the fuel efficiency of Toyota's smaller cars. The only way to get there was to use a battery, but pure electric vehicles had proven impractical. The hybrid gas–electric engine was born. The battery in these vehicles does not need to be plugged in—it gets its power from the energy normally wasted during braking. The result of this decade-long push for new technology, the Toyota Prius, has been a giant success.

Customers not only pay a price premium for the Prius, they wait months to buy it. The Prius represents what business school professors Chan Kim and Renée Mauborgne call value innovation: where a product is so new, different, and unique that customers believe there is no substitute. For many Prius buyers, a Chevy Malibu or Honda Accord just won't do, even though they can it drive it off the lot. The Prius has, as Kim and Mauborgne might say, made the competition irrelevant. In effect, "hybrid" is a new category of personal transport separate from "car."

But the Prius has done far more for Toyota than create a small niche phenomenon. The company has used what it learned from the ten-year journey to increase its speed to market on new models and improve production processes—an amazing feat for what's widely considered the world's leanest manufacturer. More than that, the Prius has created a halo around Toyota, making its vehicles hot across the board. While Detroit flounders, Toyota is making money hand over fist.

Toyota has risen fast to become the world's number two car manufacturer, and for good reason. The company exudes excellence from every pore. But a central part of the story turns on environmental innovation driving the company's vision of the marketplace. Toyota saw the Green Wave coming and responded. The company promoted value innovation and ended up with a breakthrough product that enhanced profits and sustained shareholder value. That's what Eco-Advantage is all about.

"Servicizing"

Energy guru Amory Lovins likes to say that people want cold beer and hot showers, but they don't really care how the refrigerator works or what makes the water hot. Understand this customer reality, and you open up one interesting path to Eco-Advantage.

By offering a service instead of a product, a company profits by reducing its use of materials and energy, and providing that service at the lowest cost possible. Lovins argues, for instance, that air conditioner manufacturers should offer cooling as a service—not AC units as a product—so they'd have an incentive to make the systems highly energy efficient. In some green business circles, the idea of a recasting a product as a service, often called "servicizing," is the holy grail of environmental innovation.

Some path-breaking companies are stepping up to the challenge and serving ultimate customer needs without regard to traditional product definitions. Connecticut-based chemical distributor Hubbard Hall faced a serious "disintermediation" challenge as the Internet made it easier for its customers to buy chemicals directly from the manufacturers. So the company servicized its product, offering to track chemical inventories for its customers, handle regulatory paper work, resupply them as needed, and take away empty containers. This new business model saved customers money by reducing their management time and compliance costs. It also preserved Hubbard Hall's place in the market and allowed the company to raise margins.

Trouble is, servicizing doesn't always work. Interface Flooring's attempt to lease carpet under the Evergreen brand fell flat. The business model seemed attractive. Interface would provide and maintain a company's flooring, replacing (and recycling) old carpet tiles as needed. The potential environmental gain was clear. Interface would have an incentive to make its carpet as durable as possible and get an opportunity to recycle its own product, saving energy and natural resources.

But as Chairman Ray Anderson told us, things didn't work out as planned. Tax and accounting rules, it turns out, favor sales over leases. Moreover, in most companies, a carpet purchase comes out of the capital budget, while a lease comes from the operating budget. And different people generally manage these two budgets. In short, moving expenses from the balance sheet to the income sheet is not something companies want to do.

So servicizing may not always pay off. But contemplating how to servicize a product with an eye toward reducing environmental impacts can be an illuminating exercise. What does the consumer really want from the item you sell? Here again, all the regular business cautions apply: Is there a market? How will customers react? Do they want to own the product for a good reason? Can we persuade them otherwise? Will it really lower our costs or our footprint?

GREEN-TO-GOLD PLAY 8: INTANGIBLE VALUE— BUILD CORPORATE REPUTATION AND TRUSTED BRANDS

In our celebrity-oriented world, brands matter. As the Information Age overloads customers with product options and configurations, brands provide a short-cut for customers to identify their favorite products—and for talented workers to pick employers. The better a company does at protecting its reputation and building brand trust, the more successful it will be at gaining and maintaining competitive differentiation.

Moving "Beyond Petroleum"

In 2000, British oil giant BP embarked on a massive rebranding campaign at a reported cost of $200 million. Out with the old shield logo and in with a softer sunburst dubbed "helios." A critical part of the message was a bold statement that put BP far out in front of its competitors on environmental issues. In TV and print advertisements no one could miss, the company declared itself to be "Beyond Petroleum."

Not everyone took this pitch well. The company faced harsh criticism from some environmentalists and even some fairly humorous ripostes. An NGO report, *Don't Be Fooled 2005*, lists top ten "greenwashing" ad campaigns, with BP taking the number two spot behind Ford. Another group declared the campaign "Beyond Preposterous"—as well as Beyond Pompous, Beyond Pretension, Beyond Posturing, Beyond Presumptuous, and Beyond Propaganda. Greenpeace even gave CEO Lord John Browne an "award" for "Best Impression of an Environmentalist."

Were these criticisms justified? Yes and no. BP has achieved admirable reductions in its own greenhouse gas emissions. It's one of the world's largest providers of renewable energy products such as solar panels. But even if BP's solar business reaches its target of $1 billion in sales by 2008, at least 98 percent of the company's roughly $300 billion annual revenue stream will come from oil and gas. Bottom line: BP hasn't moved beyond petroleum just yet.

What was a stodgy old oil company trying to accomplish with these ads, anyway? Did they enter into this campaign lightly? "The brand was very, very carefully positioned," said Chris Mottershead, Senior Advisor and a key player in crafting the speeches that Browne gives to declare major policy shifts. "It took a long time and lots of resources to get to the helios design and overall positioning. . . . these were deeply conscious thoughts and it was a profound, long, painful process," he added. The point of the campaign was to say something revealing about what BP was all about and to communicate to all stakeholders the new general direction of the company.

Mottershead put it in context:

> You're telling people what you think the future will be and your role in that future. Why do people pull into a BP station versus an Exxon one? Because it's saying something about their aspirations and expectations for the future. It's not that the fuel that they're buying is any better. . . . it's not like Coke and Pepsi where there's actually a difference in flavor. This has to do with a statement that tells everyone—employees, government, civil society, and some consumers—what you stand for.

In the short run, BP took more than a few hits. Wisely, the company pulled a number of advertisements and dialed back its rhetoric. The tagline was refocused on the much more defensible phrase, "It's a start." But in the longer run, BP accomplished all it could have wished for and more. Despite being in a business with large environmental impacts, the company is now seen as green. Indeed, BP comes out at the very top of our WaveRiders ranking, and Lord Browne has been on *Management Today*'s list of most admired CEOs for five years running.

Here's the real proof, though. BP's brand value, as measured by experts in measuring intangibles, has jumped significantly. A recent study of brand strength highlighted ten products with the greatest increases in brand value in recent years. Sorted by total brand value added, they are: Google, BP, Subway, iPod, DeWalt, Sony Cyber-Shot, LeapFrog, Gerber, Sierra Mist, and Eggo. BP is second only to Google, a once-in-a-generation success, and ahead of iPod, one of the greatest consumer product launches in history. BP, the study said, gained over $3 billion in brand value.

In another measure of the campaign's success, BP has found that it has become a more attractive employer for graduating engineers. Measuring the benefit exactly is impossible—the evidence is all anecdotal—but as Mottershead put it, "We don't have the recruitment problems we had ten years ago. And when I spoke to a hundred new hires, none of them worked in renewables, but every question was about green and sustainability."

If imitation is the sincerest form of flattery, BP is also doing well. Shell has a long-running ad campaign touting its green bona fides. Now, some oil and gas laggards have jumped on the ad train. Chevron launched blunt print ads warning that "the era of easy oil is

TRUTH MATTERS

Positioning a brand as environmentally friendly only works if it's true. Some companies miss this simple point: Before trying to position your products as green, make sure you have your ducks in a row. The 1980s and 1990s saw a flurry of bogus green claims. Some were just plain laughable. Hefty's highly touted biodegradable garbage bag broke down in sunlight, but not in the landfills where they would actually end up. Not good.

over" and touting its commitment to green practices. Even ExxonMobil is talking green and investing money in renewables research.

Ecomagination

More recently, GE launched its "ecomagination" campaign with an impressive list of public commitments: a doubling of investment in R&D for environmental technologies to $1.5 billion, an increase in sales of environmental products from $10 billion to $20 billion in five years, a reduction in company greenhouse gas emissions of 1 percent, while the business grows substantially. In shaping the campaign, CEO Jeff Immelt did not shy away from rigorous goals. Lorraine Bolsinger, GE's new Corporate VP dedicated solely to managing ecomagination, says, "There were five proposals for greenhouse gas emission goals—Jeff picked the toughest."

So what exactly does an ecomagination product look like? Out of the thousands of goods GE sells, just seventeen were initially selected because they improve customer operating *and* environmental performance. Some are inherently greener than the alternatives, like wind turbines and solar panels. The rest of the favored seventeen include regular products that improve on what's in the marketplace. For example, the GEnx jet engine, which will fly on new Boeing and Airbus jets, will burn 15 percent less fuel, run 30 percent quieter, emit 30 percent less nitrous oxide, and cost less to operate. Again, GE wants all ecomagination products to deliver both environmental and economic benefits to customers.

The print and TV ads for these products, and for ecomagination in general, started appearing everywhere in mid-2005. Supermodels in a coal mine pushing "cleaner coal" and dancing elephants touting "technology that's right in step with nature" helped to reposition GE as a green company.

As Bolsinger notes, "Ecomagination is carefully crafted—the 'eco' is obvious, and 'imagination' connects to our tagline 'imagination at work,' which we did immense research to get to." To those who felt the ads were over the top, Bolsinger says, "Look, part of initiative is to shine a bright light on this topic—it's purposely bold."

The company is creating intangible value by building trust in GE's brands. Along the way, it has done its homework. The company carefully vetted its marketing claims by developing "scorecards" that assessed the environmental strengths and weaknesses of the seventeen products to be promoted under the ecomagination banner. Having the data to back up the claims was a smart move and one that, so far, has saved GE from some of the complaints BP received about overreaching.

With the focus on specific products, arguably, ecomagination is as much a product play as a committed effort to go green: GE wants to *sell* those jet engines, not just have environmentalists admire them. But note how the campaign is being waged. To reach a jet engine buyer, you develop relationships with only two companies, Boeing and Airbus. You don't need to run ads in national magazines. Clearly ecomagination is much more than a product play. It's also an image advertising campaign meant to reposition GE as a company that provides solutions to society's environmental problems.

UNILEVER'S "VITALITY" POSITIONING

The line between green image marketing and green product marketing can be blurry. Unilever, for example, has made one of its new major strategic thrusts a focus on what it's calling "vitality." It's a broad idea encompassing freshness and a healthy lifestyle. A major part of the effort is traditional product marketing. The pitch is that frozen foods are captured when they are very fresh and highly nutritious. Through a multimedia approach, including expansive websites (in many languages), Unilever is connecting the vitality idea to its extensive work on sustainable agriculture, sustainable fisheries, and even recycling and other green operational efforts. It's a corporate-level green branding play, but a subtle one.

Is It All Worth It?

Many companies, even those with a lot to crow about, prefer to keep a low profile about their green initiatives. Stick your head up and you might get slapped for not really doing enough.

Sustainable business expert Joel Makower tells a story about discovering that Levi's was buying 2 percent of its cotton from organic farmers. When he asked the company to tell its story, executives were wary. They were concerned, and rightly so, that they would face tough questions about the other 98 percent of their cotton and why Levi's would make products with dangerous pesticides. As Makower says, many companies share this concern, not wanting to "draw unwanted attention to the unaddressed environmental challenges that pretty much all companies have."

Dangers run the gamut from tough questions to anti-image campaigns run by activists, as Bill Ford discovered a few years ago. Ford had promised to improve SUV fuel economy by 25 percent over five years. But after a few years, the company was forced to announce that it would not make that goal. Bill Ford's commitment to the environmental cause can't be questioned. He's been out in front on a number of issues and is trying to help Ford find the gold in green. But having taken a high-profile stand on the need for greater corporate environmental stewardship, and then missing the targets he set for the company that bears his name, Ford ended up scorned by many in the environmental community.

The Bluewater Network, an environmental NGO, ran full page ads in the *New York Times* comparing Ford to Pinocchio and declaring, "Don't buy Bill Ford's environmental promises. Don't buy his cars." Surprisingly, in late 2005, Ford went back to the green well with a new set of ads touting the environment and innovation and starring . . . Bill Ford. In this latest image campaign, he makes a new set of commitments—such as selling 250,000 hybrids per year by 2010. NGOs will be watching to see if he delivers this time.

Let's face it: Tying a brand to environmental virtue can be dangerous. The further a company is from really being green, the larger the effort required and the greater the risk entailed. But if the campaign is done right, the payoff can be substantial. A strong brand with a full "trust bank" is a valuable asset.

A deserved reputation for environmental care "inoculates" a company when bad things happen. BP's reputation, for example, could have been badly hurt by several recent incidents, including multiple explosions (and fatalities) at a Texas refinery and a 267,000 gallon oil spill in Alaska. It wasn't, though. Thanks to its positive eco-reputation, the company was given extra leeway. As one knowledge-

DAMAGING A TRUST BANK: THE DOWNSIDE IN PUMPING UP THE UPSIDE

Companies that seek Eco-Advantage through enhanced intangible value simultaneously create exposure. They must live their values every day. If the facts don't square with green claims, charges of greenwashing are sure to follow. Sometimes violations of trust may be unintentional. The Body Shop had to retract its claim that its ingredients were free of animal testing after learning that some of its suppliers bought materials from companies that didn't meet the standard. A legitimate—even innocent—mistake on the part of The Body Shop, but it shows the importance of digging into the full value chain to find hidden issues. Intentional or not, trust is built up slowly, but can be lost quickly.

able observer noted, "It was fascinating how much slack the environmental community cut BP. Their investment in being seen as good guys paid off handsomely. If ExxonMobil had done the same thing, there would have been hell to pay."

And of course the upside can be highly profitable. Companies with higher brand values have market power. They command higher prices, sell more, and develop closer relationships with customers and employees.

MAKING THE UPSIDE A CORE FOCUS: DUPONT'S SUSTAINABLE GROWTH

Cutting costs is about making business more efficient. In contrast, increasing revenue is about growth. Both sets of Green-to-Gold Plays are valuable, but a cost focus is mainly tactical. A revenue focus is often on the bigger picture and more about vision.

After more than a decade of pollution control, DuPont wanted to move the internal discussion away from only cost cutting. The company has always found it helpful to set visionary goals and make grand statements, if only to inspire employees to reach further. So the heads of the business units got together to talk about sustainability and how DuPont could play in this arena, and ended up with a new focus for the company: Sustainable Growth. With this new vision, employees would look for opportunities to drive topline growth through sustainability, with new products or a new take on old ones. The point was to add a new visionary goal to the company's lexicon and spur innovative thinking.

THE ECO-ADVANTAGE BOTTOM LINE

Driving new revenues by using an environmental focus to add value to your products, reach out to green consumers, and create new market space can produce big payoffs. Finding ways to reposition a company in the marketplace and to move the top line with environmental strategy is the cutting edge of Eco-Advantage. But keep in mind six lessons:

- Meet customer needs that actually exist.
- Don't ignore the customer's nonenvironmental needs.
- Control costs.
- Remember that green attributes rarely can stand alone: the environmental story is the second or third "button."
- Market to different niches differently.
- Don't expect a price premium.

Marketing a company's overall greenness only works well if the company has the substance to back up its green image campaign. But the intangible value generated can be substantial.

Part Three What WaveRiders Do

Understanding the Green-to-Gold Plays is important. Getting into the game, however, requires more than just a playbook. We've found that a successful journey toward Eco-Advantage starts with the right mindset and a focus on driving environmental thinking deep into corporate strategy. Our research has identified five ways of thinking that help businesses find opportunities to seize a competitive edge. We take up these approaches to developing an Eco-Advantage Mindset in Chapter Six.

Setting out on a journey with nothing to guide you but the right frame of mind is like trying to sail to a distant port with a good ship but no crew or charts. So we also provide the tools businesses need to get beyond good intentions and turn environmental focus into competitive advantage.

Chapter Seven shows you how to map environmental performance. With the right information, companies can understand how big environmental issues affect their value chain and the competitive playing field. We call this **Eco-Tracking.**

Eco-Advantage Toolkit

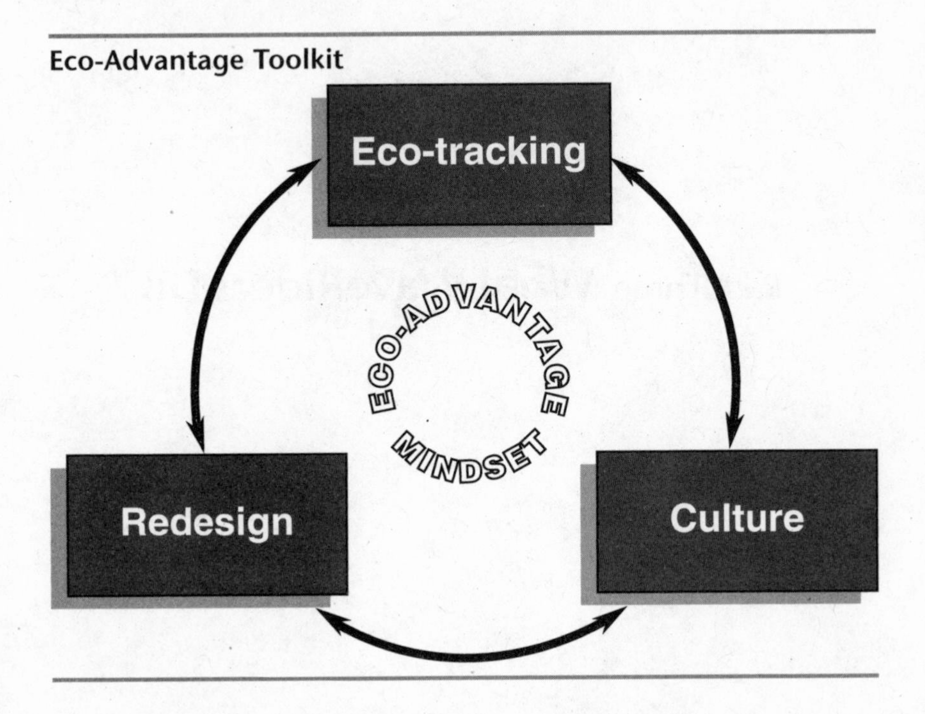

Seizing opportunities to lower cost and risk, or raise revenues and intangible value, often means redesigning both products and processes. It also means helping suppliers and customers change course and lower *their* impacts. In Chapter Eight, we discuss tools that help companies **Redesign** their entire value chains.

Finally, in Chapter Nine, we explore how to build an Eco-Advantage **Culture** and engage executives, managers, and employees in the vision. Inspiring workers with goals, ownership, and incentives that build environmental thinking into all levels of the organization will help turn environmental challenges into opportunities for profit.

With the Eco-Advantage Mindset at its core, these three tool sets make up the Eco-Advantage Toolkit (see figure), which is the basis for meaningful action and successful execution of the Green-to-Gold Plays. Some executives even talk about the importance of embedding environmental thinking and action in their "corporate DNA." We agree, and the following chapters show the way.

Chapter 6 The Eco-Advantage Mindset

In 1963, Oregon teen Dick Fosbury was a good high jumper on his high school track team, but he was nowhere near ready to compete on an international level. Five years later, Fosbury emerged as the world's best high jumper.

For decades, high jumpers had jumped over the bar using basically one method: run at it face forward, kick one leg over, and then the other. Thousands of coaches were teaching millions of kids the "proper" way to jump, working to perfect what Fosbury would show was a suboptimal method. Instead of using this traditional "straddle" or "scissor" approach, Fosbury saw a new way to clear the bar. As he approached the hurdle, he turned his back to it, arched over, and kept his legs together. This simple innovation changed the sport of high jumping forever.

Fosbury won the gold medal in the 1968 Olympic Games, breaking both the U.S. and Olympic records. In four short years, by the 1972 games, twenty-four of forty Olympic high jumpers had adopted his style. All but two medalists since 1968 have used what is now called the Fosbury Flop.

The potential to remake the sport was always there. After all, the mechanics of Fosbury's jump were easy to grasp. But only an innovative thinker, a future engineer, saw the possibility for transformation. As Fosbury told *Sports Illustrated*, "I've never tried to be a nonconformist. I just find different solutions. I'm a problem solver. That's what engineers do."

Everybody talks about out-of-the-box thinking and paradigm shifts, but it's discouraging how rarely we witness real, discontinuous change. Fosbury found an instant competitive advantage from looking at an old problem in a new way. Similarly, in the corporate world, a handful of companies are developing new ways of approaching a thorny problem: How do we grow and prosper while decreasing pollution and conserving natural resources?

WaveRiders build a foundation for Eco-Advantage by reframing how everyone in the company looks at environmental issues. For these companies, environmental thinking is not always the final word on strategy, but it *is* always a consideration.

In our research, we've found that this new mindset is absolutely critical to managing eco-risks, driving innovation, and turning environmental pressures into competitive advantage. This chapter highlights how WaveRiders use an environmental lens to change the way they think and sharpen their business strategies. After a while, these companies don't have to focus consciously on finding an alternative perspective. Environmental thinking becomes intrinsic to how they do business. Deeply embedded, the **Eco-Advantage Mindset** arises naturally at every opportunity.

Some basic rules to get you there:

- **Look at the forest, not the trees.** WaveRiders think broadly about (a) the time frames involved in investment and strategy decisions, (b) the full range of potential payoffs from those investments, including hard-to-measure intangible gains, and (c) possibilities for adding value across the full chain of production.
- **Start at the top.** Every company we found leveraging Eco-Advantage had a commitment to environmental thinking at the very top of the organization.

- **Adopt the Apollo 13 principle—"No" is not an option.** In leading companies, management gives the organization bold environmental goals and seemingly impossible tasks—and refuses to accept failure.
- **Recognize that feelings are facts.** Top performers know that what NGOs, employees, customers, communities, and other stakeholders *feel* about a company's environmental performance and reputation can be much more important than the reality.
- **Do the right thing.** We've seen time and time again that WaveRiders make choices based on core values, including caring for the environment, even when it might not pay off in the short run.

LOOK AT THE FOREST, NOT THE TREES

If Dick Fosbury had stuck with traditional high-jumping techniques, he never would have won an Olympic gold medal. Fosbury, though, looked at the big issue first (getting over the bar) and then reasoned backwards to find the best way to do it. By concentrating on the forest, he found the strategic advantage. WaveRiders in the corporate world do the same. True, many business books push broad thinking, but taking up the environmental lens requires stretching the mind in new directions. Nature, after all, cuts a very wide swath.

In factoring environmental considerations into their strategic thinking, WaveRiders broaden their vision across three critical dimensions. They consider issues in both short- and long-time horizons. They calculate payoffs more broadly than others and are more attuned to intangible costs and benefits. They don't let the traditional boundaries of their business limit their vision, and they search for ways to improve performance throughout their value chains.

Time: The Strategic Term

WaveRiders consider short-term financial impacts, but they look past quarterly financial results before making important decisions. They know that maximizing shareholder value is not the same thing as maximizing quarterly profits. And they recognize that proper analysis of some issues, including many environmental challenges, requires longer timelines.

Companies make long-term business decisions all the time. They spend millions on R&D when the potential payoffs down the road are, at best, uncertain. They enter tenuous new markets like China and India in the hope that business will boom. And they invest in leadership training to build up "bench strength" and prepare future executives. The Eco-Advantage Mindset requires that companies bring the same long-term perspective to environmental strategy.

Rick Paulson is the Plant Manager for Intel's multibillion-dollar chip fabrication plant near Phoenix, Arizona. As the head of "Fab 22," he's responsible for producing Intel's latest chips in vast quantities using some of the most advanced technology in the world. And he has to do it quickly, cheaply, safely, and profitably. But like all Intel executives, Paulson must think beyond today's operations. Intel lives or dies by its ability to build the *next* chip plant—and the next chip. Fab 22 will be outdated very soon. So Paulson has an eye on Intel's future production needs. He knows that he must keep community groups, activists, and regulators happy for one simple reason: Their unhappiness can easily slow or block expansion plans, costing the company millions of dollars. No wonder Paulson talks about making decisions not for the short term, but for the "strategic term."

How long is the strategic term? It depends on the business. It may be a year or two, or much further out. In the late 1960s, executives at Royal Dutch/Shell began looking for ways to prepare for an increasingly unstable oil market. The result was a planning group that focused on painting pictures of possible futures—scenarios that would help the company think about what its business might look like over the long term. Among other futures the team famously imagined were the rise of OPEC and the fall of the Soviet Union.

More recently, scenarios have helped Shell set environmental strategy. We talked with the head of the group, Albert Bressand, and a group of other senior executives. Over a lunch in headquarters in The Hague, Netherlands, they explained how scenarios help them devise better strategies.

"First, you should know that scenarios are *not* forecasts," Lex Holst, VP of Health, Safety, and Environment, said, "but consistent pictures of a possible future." Added Mark Weintraub, the executive who writes the annual Shell Report, "We ask ourselves, 'Will 20 percent of all new cars in 2020 be powered by hydrogen fuel cells?' We don't know, but we can hold up a strategy to this picture and ask, 'Is my plan robust in this kind of world?' "

Will the world move toward increased energy efficiency and demand much more renewable energy? If so, Shell is facing company-altering, perhaps company-threatening, changes. So it's worth thinking long-term, asking tough questions, and building pictures of the future. These kinds of scenarios helped Shell decide to move into the hydrogen business in the first place.

Other companies are talking more informally but still thinking long-term. Executives at IKEA are exploring how their business might greatly reduce its dependence on fossil fuels. Given today's infrastructure and technologies, could a furniture business consume *no* oil or gas? It might seem farfetched. But IKEA executives know that the world is changing and they want to be ready to move proactively into the energy future one step ahead of the competition. Even smart fossil-fuel suppliers like BP and Shell are planning for a day with much less fossil fuel in the mix.

What long-term environmental pressures could sink your business? And which of them might offer opportunities for growth? Until you ask those questions in a serious and systematic way, the future will control you instead of the other way around. Big difference.

So WaveRiders think broadly about time. Great, you might say, but taking the long view can create seemingly very tough choices now. And the market can be quite unforgiving in the short term. The trade-offs can look especially ill advised if you consider only the financial costs. Changing or retooling longstanding production processes or reformulating successful products often costs a bundle of

NOTABLE QUOTE: THE GRANDCHILDREN TEST

For Jim Rogers, CEO of energy company Cinergy, the long run extends a generation or two. When asked why a company that makes its money burning coal is thinking about the environment and climate change, Rogers says, "I apply what I call the grandchildren test. Simply put, when my grandchildren get to be my age, will they say that their granddaddy made good decisions that remain good decisions?"

money upfront. But not paying attention to changing circumstances and new pressures, including environmental ones, can hurt a business far more.

Payoffs: Beyond the Immediate and Obvious

Every business faces countless decisions about how to invest the next dollar. Should we spend more money on R&D, on new equipment, or on a new marketing campaign? Every business has some process, either formal or informal, for making these cost–benefit calculations and coming to a conclusion. For most companies, the decision hinges on out-of-pocket costs and potential financial returns.

WaveRiders operate like all other companies, but they think differently. They include more than the obvious dollar payoff in their decision-making. When considering the return on an investment, they factor in benefits such as enhanced brand image and corporate reputation, improved employee morale, community support, reduced governmental red tape, increased speed to market, and competitive differentiation. These intangibles are hard to measure, but smart companies include them in their strategic planning, despite the difficulty. Leading-edge managers have taught themselves to fold intangibles into their calculations at every turn because they know that immeasurables sometimes produce the greatest value.

We heard one story from industrial giant 3M that really impressed us. No company knows more about the beauty of small things than 3M, which has seen its little yellow sticky notes grow into a billion-dollar product. The creation of the first Post-it is a famous case of serendipity, but the growth of the brand into a mainstay of the company's balance sheet didn't just happen. The company has always had a remarkable eye for product extension opportunities. This isn't a story about that, though. Instead, it's a tale about revenues foregone, values stuck to, follow-through on promises made . . . and thinking broadly about what the real payoff of a decision is—a key element of the Eco-Advantage Mindset.

3M product managers had identified a new and fast-growing market for Post-its. Consumers wanted to put notes on vertical surfaces like computer monitors, but as millions of people were annoyed to find out, Post-its tend to fall off in these situations. The simple solution was to design a stronger adhesive, which the world-class scientists at 3M quickly did. But that led to a big internal problem.

The new adhesive required the use of industrial solvents that release dangerous pollutants called volatile organic compounds (VOCs). These toxic chemicals bring with them a host of problems including air pollution, worker safety issues, and potential liability. To avoid these problems, 3M CEO Livio DeSimone had declared in the early 1990s that no new investments would be made in technologies that produced VOCs. No exceptions.

That mandate left brand managers with a potential blockbuster product that they couldn't bring to market. They asked the researchers to find an adhesive that worked without using VOC-producing solvents. Six long years later, in 2003, 3M launched Super Sticky Notes to much success and millions of dollars in sales.

3M is keeping the opportunity costs of this six-year wait close to the vest, but based on the size of the Post-its business, we believe 3M gave up tens of millions of dollars in revenues to stick by the no-solvents pledge. Yet when asked today if the choice was worth it, 3M executives say "absolutely." And this is partly because they look at their payoff matrix differently from executives in other companies.

In making this decision, 3M calculated what the company calls the "total cost" of using solvents, which includes some hard-to-measure but important expenses. By sticking to its no-solvents pledge—by thinking in the strategic term instead of the short term—3M narrowed its exposure to new air pollution regulations, cut monitoring and compliance costs, and reduced the risk of EPA or state fines. Simultaneously, 3M showed its customers, local communities, regulators, and NGOs that it was serious about reducing its ecological footprint. The decision also made clear to workers that 3M puts safety ahead of profits, a message that pays big dividends in employee dedication.

As 3M Vice President Kathy Reed tells it, with these intangible benefits included, the total cost of *not* going the solvent-less route was too high. Of course DeSimone could not have known that the no solvents rule would result in a six-year delay in sales, but he certainly knew there would be trade-offs.

In the end, finding a way around the solvent roadblock also turned out to be good for the long-term financial bottom line. As Post-it

GREEN COST ACCOUNTING

In most companies, nobody really knows how much environmental issues cost the business, or how much a change in practices helps the bottom line. The problem is that the relevant costs may be spread out among many different departments or buried in "general and administrative" accounts. A number of companies have worked to unveil these costs, which are normally hidden from management scrutiny. Breaking out environmental spending separately helps to clarify the fully loaded costs of the company's products or processes.

The environmental group at Northeast Utilities has gone a step further. Whenever they help those in line operations address an environmental issue, they estimate the savings achieved not only in direct costs, but also in management time reduced, regulatory compliance burden lifted, and other indirect costs avoided. At the end of the year, they tabulate these savings and issue a report they call "Earning Our Keep," highlighting their value to the business.

plant manager Valerie Young noted, "solvent-less technology is half the cost and two times the speed." Avoiding all the special handling required of the toxic solvents sped up operations and removed some serious health risks.

The point, though, is that no one at 3M could have promised at the time that a solvent-less technology would ultimately be cheaper. Developing the new non-solvent technologies and changing the manufacturing process looked mainly like an expense—one the CEO demanded, but an expense nevertheless. Looking solely at the bottom line, 3M would not have delayed the release of the Super Sticky line. And without thinking beyond the financial cost and considering other softer paybacks, 3M *would never have found a better way to go.*

The "trees" that 3M could easily have gotten lost in were the immediate revenues and profits from launching the Super Sticky line. The "forest" was a stronger business that was safer for employees, better for the environment, and more profitable.

In some cases, seeing the forest can be literal. Smart companies in the natural resource realm (forestry, oil, and mining) are now taking into account how their actions affect wilderness and the plants and animals living there, or more scientifically, biodiversity. Almost by definition, extractive industries destroy biodiversity. Cutting trees or sweeping away the land to get to minerals or oil can demolish the local environment.

The word "green" probably doesn't leap to mind when thinking about extractive industries, so we were more than a little surprised when we asked environmental advocates to name some companies handling biodiversity issues well. Glenn Prickett, the Director of the Center for Environmental Leadership and Business within the global NGO Conservation International, highlighted the work of mining giant Rio Tinto. As Prickett said, "Twenty years ago, an NGO formed just to go after Rio Tinto, so they've felt the pain. Now Rio Tinto has the most sophisticated biodiversity strategy out there."

Rio Tinto needs access to land for exploration if it wants to continue satisfying the world's growing demand for minerals and metals like copper, aluminum, and iron. But permission to mine is getting much harder to obtain as open land disappears. As Mark Twain once said, "Buy land. They've stopped making it." Property owners, communities, native peoples, and local governments today have less interest in their land being sliced open, even for substantial economic gains. They require a show of good faith from any company that promises riches.

To find ways to please the communities that control the land, Rio Tinto became, of necessity, a good caretaker of biodiversity. But its commitment went beyond expediency. The company shared with us its internal guidance document, "Sustaining a Natural Balance," which lays out why they care about biodiversity and how operational executives should manage it. This impressive manual describes a five-step process for assessing potential exploration tracts, working with communities and NGOs, and finding ways to mitigate as much of the damage caused by extraction as possible. Rio Tinto will never please all its critics, or even fully undo the damage from mining, but by working with stakeholders, the company is moving toward a better balance of economic growth and environmental protection.

Once considered merely nice to have, environmental plans and community engagement strategies have become must-have items. As Dave Richards, Rio Tinto's principal adviser for environmental issues, told us, "Our license to operate is granted not just by the community at the plant gate, but society at large. If you play hardball with those interests, you'll have a hard life. This is a conscious attempt to change a risk into an opportunity."

Boundaries: Beyond the Factory Gates

Before its "cadmium crisis," Sony probably did not consider an obscure Chinese supplier of control wires a part of its inner circle. But once you strand 1.3 million game systems in a Dutch warehouse a month before Christmas, your definition of "family" gets much bigger.

Thinking of your company as just your own factories, offices, and other hard assets is outmoded. Corporations operate within a global production network, with suppliers plugging into various parts of the process. The boundaries between us and them have blurred. Intangibles like brand value and employee knowledge and skills are part of every company's value, the gold WaveRiders get from being green. These "soft" assets are assets just the same and need to be tended with great care.

When Doug Daft invited Dan Esty to join his newly created Coca-Cola Environmental Advisory Board, he explained his vision for the new group in simple terms: With a market capitalization of $115 billion, Coca-Cola's book value amounted to only about $15 billion, leaving a whopping $100 billion in intangible value. Daft understood, as does his successor Neville Isdell, that when you are the guardian of the world's most valuable brand, environmental mistakes can cost millions, even billions of dollars.

It's no longer enough for companies to say "we do our part for the planet." That was last century, and those were too often empty words. We live today in a world of extended producer responsibility. Companies cannot dodge the environmental problems up and down their value chains, from the furthest upstream supplier to the most remote downstream customer. Simply put, what happens *outside* your four walls is often what counts most.

This reality requires asking new questions: Are workers at your suppliers' factories facing toxic exposures or other unhealthy conditions? Do your suppliers dump hazardous waste in the local river?

Do your customers discard your product in a manner that causes litter or pollution? These questions and many more are out on the table now.

The wonderful and horrible thing about the Internet is that information is everywhere. One story of suspected child labor or of dumping toxic waste in a river anywhere in the world will instantly spread, tarnishing your brand and destroying value. Thinking about the environmental effects of your products and production processes beyond your own factory gates isn't just advisable, it's essential.

Low-cost furnishings champ IKEA has committed immense resources to tracking where all its products come from, particularly the wood. The company has set aggressive goals: It will buy no wood from areas of high conservation value, accept no lumber from illegal logging, and move toward purchasing only wood certified as sustainably harvested.

Is IKEA taking too much responsibility for what happens outside the confines of its business? After all, the company does not purchase wood directly; it buys the furniture made with the wood. But that's not how IKEA's top brass see it. They believe their customers expect nothing less from them.

Sitting in a light, open IKEA office in tiny Gelterkinden, Switzerland, Chief Forester Gudmond Vollbrecht told us how his team creates value. "My foresters [eighteen in all] travel up to 140 days per year to see furniture and wood suppliers first-hand." This expensive and extensive effort isn't just about doing the right thing, Vollbrecht said. "If we can't find business value and be part of the business agenda, we'll fail. We've found that knowing our supply chain is great business mapping also—we just know our business better." Thanks to its tracking efforts, IKEA is finding new efficiencies and ways to cut out the middle man. For a company that prides itself on delivering reasonably priced goods to eager customers, improvements that contain costs are big strategic victories.

Just as IKEA has learned that its upstream supply chain can be more than just a source of risk, so other companies are finding new opportunities downstream. HP, for example, saw that its customers had trouble getting rid of old toner cartridges for printers. The company also noticed sales slipping as new companies sprang up to sell reconditioned cartridges. Instead of walking away and leaving the customer to deal with this disposal problem, HP launched its own very successful recycling and remanufacturing business.

The printer business is HP's crown jewel, and remanufactured toner cartridges are now a high-margin, hundred-million-dollar business. Some 11 million cartridges are now reused annually, and over 80 million have been recycled since 1991. More importantly, if HP had not built this Planet Partners reuse business, someone else would have. Initially, thinking about the full life cycle of its product was a defensive move, but this defense also proved to be a good offense—a major upside success story with a highly profitable ending.

IKEA, HP, and other smart companies are looking for opportunities across the value chain. This life-cycle way of thinking yields valuable insights on products and production processes. It also helps managers understand their businesses better and capture value wherever they might find it in the value chain.

START AT THE TOP

In many ways, seeking Eco-Advantage requires a broad-based effort involving all levels of a company. But real environmental leadership has to start at the top. In every WaveRider we identified, executives told us that commitment from the highest levels was a critical element of success—the only way to engage middle management and employees down the line in the challenge of remaking a business to be more environmentally sound. Why would a BP refinery manager work to cut emissions without a "request" from the CEO? How could 3M product managers have waited for years to yield profits from the Super Sticky Post-its if the directive had not come from the top of the organizational chart?

In all of our WaveRiders, the CEO and other senior executives have engaged on business sustainability issues, sometimes very personally. At DuPont, Chad Holliday serves not only as the chairman and CEO but, by his own declaration, as Chief Safety and Environmental Officer—a title that sends a powerful message through the entire company. Holliday also chaired the World Business Council for Sustainable Development. And he coauthored *Walking the Talk*, a book that makes the business case for sustainable development.

STEPPING UP TO THE CHALLENGE

We've already told the story of how BP CEO Lord John Browne vowed to take his company "beyond petroleum" and how that seemingly rash promise has now translated into $1.5 billion in savings. His pledge came after his bold public acceptance of climate change as a real and pressing issue—well ahead of his industry peers. In contrast, former ExxonMobil CEO Lee Raymond remained a vocal doubter of the science on global warming. Today, Browne has assured BP what he's called "a seat at the table" at just about every serious conversation on environmental regulation and the future of energy policy. And as Lee Raymond sails into retirement, his company will almost certainly change its tune. The price—both in access to government and public perception—for being combatively anti-environmental is rising.

Maybe no CEO has pushed for sustainability harder than Julio Moura, head of GrupoNueva, a $2.4-billion conglomerate with operations in Central and South America. The company owns, among other things, a forestry business and a manufacturing operation that makes polyvinyl chloride (PVC) piping for the construction industry—enterprises that seem the antithesis of green. Yet the company was founded to think green, and at Moura's insistence, his top executives have worked overtime to address environmental problems.

One of GrupoNueva's top priorities was eliminating the use of lead-based stabilizers in the production of PVC. Since the problem originated upstream in the supply chain, the company set out to solve it on site. "Our chemists spent one year working with suppliers and came up with a calcium-zinc solution," Maria Emilia Correa, VP of Social and Environmental Responsibility, told us. "We set a goal that the cost of production would not go up more than 1 percent . . . and they pulled it off!"

The success, though, was temporary. When it became clear that GrupoNueva couldn't guarantee suppliers enough volume to make the production change, the leadership team came up with a novel solution: share their thinking with competitors. Moura hit the road, telling other PVC pipe buyers about what lead does to the workers and the children living near their suppliers in Peru. He asked them to join with him to guarantee suppliers enough volume to justify switching to the calcium-zinc technology. And it worked. What better example could there be of "walking the talk" than that?

ADOPT THE APOLLO 13 PRINCIPLE—"NO" IS NOT AN OPTION

"Houston, we've had a problem here."

Apollo 13's April 1970 lunar mission was mostly a ho-hum event—men had already walked on the moon twice. But then an oxygen tank exploded 200,000 miles from Earth, and astronaut Jack Swigert radioed back his memorably understated message. That left NASA flight director Gene Kranz with the seemingly impossible task of bringing Swigert, mission commander Jim Lovell, and astronaut Fred Haise back home alive—a four-day trip in a two-man lunar module with carbon dioxide filters that couldn't handle the load.

If you saw the Tom Hanks movie, you're likely to remember Kranz's charge to his team of engineers. Find a way, he told them, using only what's available to the astronauts, to filter the air, or they will die before we can get them home. As Kranz put it, "Failure is not an option"—a dictum that has become a modern classic. And it worked. Facing inflexible constraints, the engineers innovated, and while the world watched anxiously, the three astronauts made it safely home.

In business, the stakes are rarely so stark, but the greenest companies also don't take no for an answer. Like Gene Kranz, WaveRiders place very tough, seemingly unachievable demands on their organizations and workers, and they hold their feet to the fire until the goals are met.

In the early 1990s, after the new national Toxics Release Inventory had pegged it as America's—and arguably the world's—leading polluter, DuPont started to get religion about going green. Determined not to be "number one" in such an unwanted category, Chad Holliday's predecessor as CEO, Ed Woolard, sent a clear message to the business unit: Reduce hazardous waste. Woolard was serious about it—he set a company-wide goal of "Zero Waste." Business units would soon find out just how serious he was.

One huge facility in Victoria, Texas, proved to be a test case. Opened in 1951, the Victoria plant specializes in nylon intermediates, products that are used to make other products. The intermediates are a major component of Stainmaster carpets, Lycra Spandex, and everything from luggage to seatbelts to panty hose. Unfortunately, manufacturing them is dirty work. In 1990, the Victoria plant created 35 million pounds of toxic waste—bad stuff like benzene and sulfuric acid, much of which was put into deep wells.

As an intermediate product in a long value chain, the business operates on tight margins. That was problem one. Problem two was that cleaning up the mess was, it seemed, going to be expensive, far beyond what the tight margins allowed. Line executives came to Woolard and said that significantly reducing hazardous waste would cost $500 million. His response was simple: "Wrong answer." So they reworked the plan and said it would cost $200 million. His response again: "Wrong answer."

In the end, the net present cost to change the Victoria business was almost zero. The engineers found ways to reduce the use of toxics at the source by changing processes. They worked with the local community to build wetlands around the facility to help clean the water naturally. And they sold some of the by-products that had once been waste. By 2002, Victoria released less than 10 million pounds of toxics, a reduction of over 70 percent, even while production grew.

That's what can happen when you put smart business people in a room and tell them they can't leave until they have an answer that protects profits *and* serves the environment.

WaveRiders are tough on everyone, including suppliers. When Dell wanted to print its product catalogs on recycled paper, the paper companies said they couldn't do it, but Dell refused to budge, and its catalogs are now printed on 10 percent recycled content. When FedEx Kinko's and Timberland wanted to use renewable energy at some of their facilities, both companies demanded and got price parity from renewable energy suppliers, at least over the period of the initial contract. When companies ask the seemingly impossible of suppliers, often the first reaction is no. But WaveRiders keep at it and demand a solution.

By the way, things haven't changed much at DuPont since it first caught the spirit of the Apollo 13 Principle. The company's energy managers are currently expected to hit a target of 10 percent of the company's energy coming from renewable sources while *saving* $8 million. They must feel like those NASA engineers.

THE APOLLO 13 PRINCIPLE

Think big. Look at your opportunities for environmental gains. Give your businesses seemingly impossible goals that reflect an ambitious environmental vision. Then step back and let the company innovate. Tell them to find a way to produce your product with less material, less energy, and less waste, but without spending any more money to do it. Success will not come easily nor every time. But the pressure can unleash amazing advances.

RECOGNIZE THAT FEELINGS ARE FACTS

Flying into Phoenix, Arizona, on a clear day can be a delight. The country is a beautiful, dry desert painted in shades of brown and red. But as you descend into Sky Harbor International airport, the vision is disrupted by splashes of rich green that don't quite fit. Those are the golf courses.

Golfers can probably pick out some of the most famous courses from five thousand feet. The Troon North Golf Club in nearby Scottsdale is, in the words of *Golf Magazine*, "built around dramatic granite boulders, striking elevation changes, and spectacular carries over desert washes." But like the many other courses that dot the area, Troon North has an appetite for water that just won't quit, in a part of the country that has little water to spare.

Golf courses in the Southwest need more water than in any other part of the country—about 88 *million* gallons per year for every single course. The water needed for just one Arizona golf course could fill 12,000 swimming pools or provide for all the water needs for about 1,500 Americans (or 20,000 Africans).

This is not a diatribe against golf courses. The point is that if you went looking for a way to reduce water use, golf courses would be a good place to start. But in Arizona, as in all places, it's easier to point a finger at industrial interests than at something you enjoy personally. Thus Intel, which has a very large facility with two semiconductor fabrication plants in Chandler, Arizona, feels the pressure. As Plant Manager Rick Paulson says, "The hottest issue here is water use."

Chip manufacturing does, in fact, require a fair amount of water. The entire Chandler site uses over 600 million gallons a year. That's a drop in the bucket compared to the 8 billion gallons used by nearby golf courses. If local golf courses were just 7 percent more efficient in their sprinkling, they could make up for *all* of Intel's water use. But having the facts on your side is never enough.

In Arizona, keeping the community happy means using water carefully, especially if you're a highly visible Fortune 100 company with a physical plant that takes up lots of space. Spending millions of dollars annually to recycle water, as Intel does at Chandler, is not an investment the company would make if it were looking only at quarterly financial results. Nor is Intel's water resource management program required by law. Intel executives feel it's the right thing to do, and by showing sensitivity to the local community's needs, they maintain Intel's license to operate.

Recently, the company received the go-ahead to build a new chip fabrication plant in just months—a process that might otherwise have taken years. How much is fast approval for a new production facility worth? Of course, it's impossible to calculate precisely, but speeding up the development of a multibillion-dollar factory is worth a lot of money.

To our way of thinking, Intel is dealing with community pressures in a very smart way. Doing the right thing and doing what's good for business are not mutually exclusive.

The perception gap is even worse with air quality concerns. Intel's Chandler site produces about the same amount of volatile organic compounds as a local gas station, yet Rick Paulson points out, "We have a problem with the perception, be it real or not, of Intel's contribution to air quality—it's not what the data is, it's what the *perception* is." Asking an engineer to ignore real data and deal with feelings is not easy.

Many other companies have fallen into the gap between the engineer's analysis and public perception. Remember how Shell conducted detailed analysis of the plan to sink the Brent Spar oil platform in the North Sea? The solution was not ideal, but it was most likely the best option. For protestors and Shell consumers around the world, though, impressions quickly trumped facts. Or consider GE's

HOW MONSANTO'S SUSTAINABILITY CRUSADE BACKFIRED

Robert Shapiro was a darling of the environmental community. During his tenure as Monsanto's CEO from 1995 to 2000, Shapiro put sustainability at the very center of his corporate strategy. He envisioned a world where innovations in life sciences, genetic engineering, and biotechnology would allow crops to grow without pesticides or fertilizers, lawns to flourish without water, and rice and other foods to provide essential vitamins. He was a visionary.

Shapiro and his team of Monsanto scientists knew they could bring about an agricultural revolution. And they were certain that they could do so safely. After all, Americans had been eating bioengineered soybeans and other crops for years, without the slightest sign of ill effects. But what Shapiro and his top management didn't know was that the European consumer wasn't at all keen on the idea of genetically modified food.

Robert Shapiro brought his GMO-based revolution to Europe with all the confidence in the world, only to unleash a firestorm of protest. With cries of "Frankenstein Food" echoing all around them, Monsanto executives tried to mount rational, science-based arguments to calm the fears of French, German, and Italian consumers, all to no avail. Before long, Monsanto was out of the European market, and the company itself was teetering on the brink. What the superanalytic Shapiro missed is that feelings are facts.

years-long battle with regulators over the proper solution to the problem of toxic PCBs in the Hudson River. Water scientists tell us that GE is basically right—leaving the chemicals where they are is not ideal, but probably the best option for reducing human exposure. But tell that to the people living along the river.

Smart companies are realizing that feelings in the marketplace are just as real to the people who have them as the company's facts are to them. Like Intel, they've seen that managing perceptions and paying sincere attention to the environmental concerns of the community can have a positive effect on the bottom line. Understanding this reality is vital to getting in the right mindset to create Eco-Advantage.

BRANDED LITTER

Big, public brands such as Coca-Cola and McDonald's get disproportionate public attention—for good and bad. Consider the problem of "branded litter." When people see trash on the road with the golden arches on it, they think, even subconsciously, "McDonald's is dumping garbage everywhere." That may not be McDonald's fault, but that's what people feel.

DO THE RIGHT THING

IKEA is a famously penny-pinching company—a habit driven by founder Ingvar Kamprad, reputed to be one of the ten richest men in the world (a rumor IKEA strongly denies). Kamprad flies coach, stays in cheap hotels, and drives the same beat-up Volvo he's had for years.

Bob Kay, the IKEA store manager in Paramus, New Jersey, told us a tale that shows just how hyper-thrifty Kamprad can be. Once when he was visiting Kay's store, the company founder saw a swept-up pile of dust and debris that included some little wooden pencils customers and employees use around the store. As Kay tells it, "Ingvar said, 'Bob, don't let them throw out the pencils!' So I got down on the floor and picked up these pencils worth less than a penny."

Given this attention to every cent (or Swedish krona) wasted, it might seem odd that IKEA spends millions of dollars on environmental initiatives. At Kay's store, for example, workers sort product packaging while receiving new inventory. They create piles for different kinds of plastic, wood, and metal, and the store pays to have the different types of waste recycled. None of this is to meet regulatory demands. What's more, the extra step slows down inventory replenishment and costs money in labor and waste management. But IKEA sees the effort as critical to its spirit of stewardship.

So why is tight-fisted IKEA spending extra money on the environment? A big part of the answer lies in a mantra posted around the offices: "Low prices—but not at any price." Those are more than words. The slogan holds real meaning to every IKEA employee we

DO YOU NEED A HISTORY OF THINKING GREEN?

Companies do not need to be founded with an environmental ethic in mind to believe that taking care of the environment is the right thing to do. Sure, some WaveRiders have talked about this for many years. Herman Miller's founder D. J. DePree, a religious man who felt a deep responsibility to the world, declared in the early 1950s that the company "will be a good corporate neighbor by being a good steward of the environment." If this was not the first mention of environmental stewardship in the corporate world, it must be fairly close.

Yes, it helps if a belief in stewardship is built into a company's history. But any company can make the commitment and find innovative ways to follow through on it. Some companies have come to these values after building a legacy that has been anything but eco-centered. Chiquita Brands, for example, has one of the most tortured corporate histories in the world. Yet today, the company's executives say that their relatively new focus on environmental and social issues is a key to the company's success.

met, and the company's actions back it up all the way to the recycling bin.

The business argument for environmental care is being recognized by more than only privately held Swedish companies. Before HP's ill-fated merger with Compaq drove her from office, CEO Carly Fiorina gave a speech about why companies should make environmental and social responsibility a core part of doing business. Three of her four themes focused on how going green was good for business. But the very first reason she gave was simply that "It's the right thing to do."

Top executives at companies such as Nike, McDonald's, and Alcan have publicly said the same thing in the same words. In his online statement of values, Herman Miller CEO Brian Walker writes, "We advocate for the environment for the simple reason that we believe protecting our fragile environment is the right thing to do."

Over and over during our research, we asked, "Why are you doing this? Why spend money on something that's good for the environment or is so intangible?" And repeatedly we heard this same mantra, often accompanied by a quizzical look, "It's the right thing to do." Skeptics might say that these corporate leaders are just mouthing empty platitudes, but we hear sincere commitment. And we've seen these same executives taking action to back their words, often at risk to the short-term bottom line.

More than anyone else, the people in the trenches know and care. If they believe that their employer is trying to be both moral and profitable, and if they see this spirit radiating down from the very top of the company in an uncompromising way, they are more likely to bring their full talents to bear. That's how you create real Eco-Advantage.

THE ECO-ADVANTAGE BOTTOM LINE

Developing the right mindset is critical to success in a Green-to-Gold world. To bring the environmental lens to your company's strategic focus:

- Look at the forest, not the trees: Think broadly about timeframes, payoffs, and boundaries.
- Start at the top: Senior executives, especially the CEO, must be engaged in setting the vision.
- Adopt the Apollo 13 Principle: Establish tough environmental goals, and don't take no for an answer.
- Recognize that feelings are facts: Emotions and perceptions carry enormous weight—and the customer is never wrong.
- Do the right thing: Clear environmental values inspire employees, customers, regulators, and potential adversaries alike.

Chapter 7 Eco-Tracking

The Eco-Advantage Mindset is a powerful motivator and the core of the environmental lens that helps companies step up to challenges and find opportunities for seizing advantage. But it's just the beginning. Companies need tools to get going. We start here with the elements of the toolkit that help companies understand where they are.

Getting the lay of the land requires thinking and analysis that might not come naturally. Eco-Tracking helps to answer fundamental but sometimes unfamiliar questions:

- What are the company's big environmental impacts?
- When and where do those impacts arise? During manufacturing? During shipping and distribution? Upstream in the supply chain? Or downstream in the hands of customers?
- How do others view the company's environmental performance?

These questions can be difficult to answer. But leading companies use a core set of Eco-Tracking tools to develop an environmental self-portrait and help them manage for Eco-Advantage. Following their lead, we suggest that you:

- Trace your environmental footprint.
- Capture data and create metrics.
- Set up environmental management systems.
- Partner for advantage.

Eco-Advantage Toolkit—Eco-Tracking

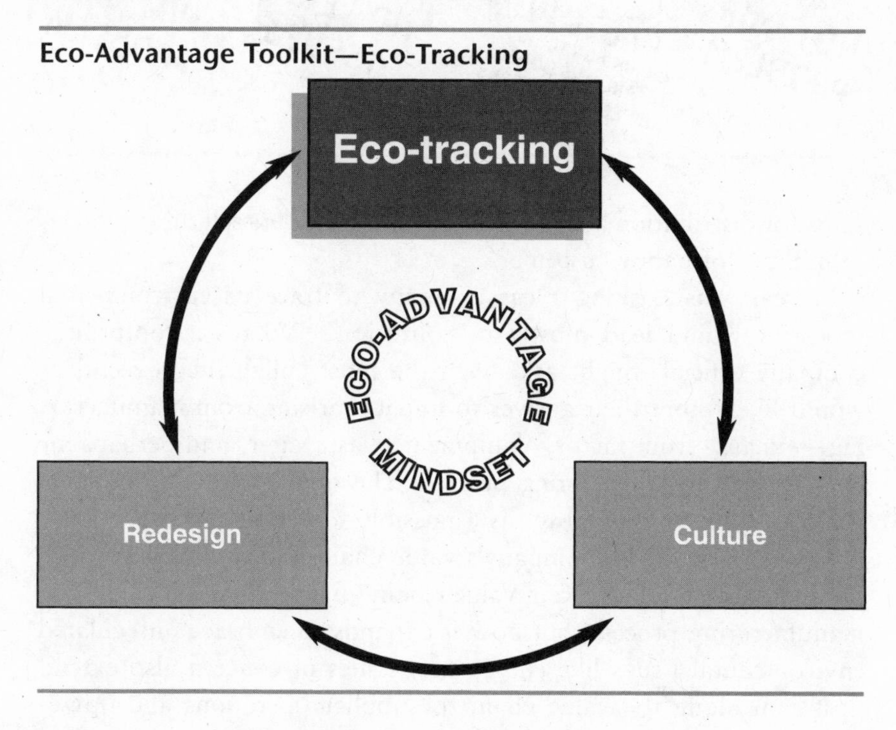

TRACE YOUR ENVIRONMENTAL FOOTPRINT

Every company leaves a mark on the world through the products it makes and services it offers. The more resources it uses or pollution it produces, the bigger its footprint.

Imagine an extremely simple version of the value chain for a car (see "Simplified Car Value Chain" figure). From the manufacturer's perspective, parts—engines, doors, windshields, seat belts, and a thousand others—come in one end of the factory. The workers assemble the vehicle, paint it, and send it out the door. Finished cars and SUVs roll out of the plant and onto trucks, trains, and container

Simplified Car Value Chain

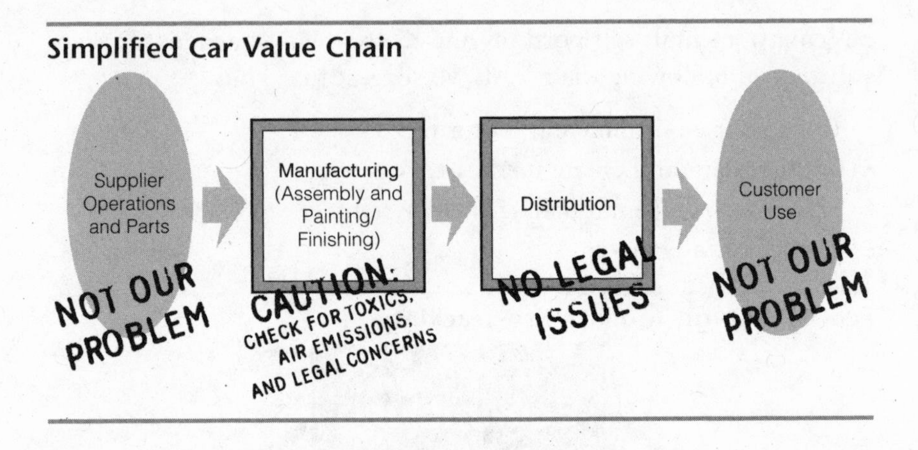

ships for distribution around the world. Car dealers sell these vehicles to millions of happy customers.

In years past, asking a car company to trace its environmental footprint would lead mostly to confusion. "What's a footprint?" company officials might ask. Even the most enlightened executives would likely limit their answer to impacts arising from manufacturing—exhaust from factory equipment, wastewater, and perhaps air emissions from the painting process. The other boxes in the value chain wouldn't even register as a possible source of concern.

The new view of a company's value chain and footprint is much broader (see "Extended Car Value Chain" figure). It still includes the manufacturing process, but now it expands to embrace unregulated environmental issues like energy use. Issues of concern also extend upstream along the value chain to supplier operations and downstream to customer use and ultimate disposal. As we've said many times, in a world of extended producer responsibility, what happens outside the factory gates now counts on your environmental balance sheet.

To get a handle on the car's real footprint, the manufacturer must ask how its suppliers operate. Where does the steel come from and how was it forged? What toxic emissions does the metal fabrication produce? Then, looking downstream: How much gas will the driver burn during the "use phase" of the product? What about the greenhouse gas that comes out the tailpipe? And what happens years later when the vehicle is hauled to the junkyard? The total footprint now reflects the car's full life cycle.

Extended Car Value Chain

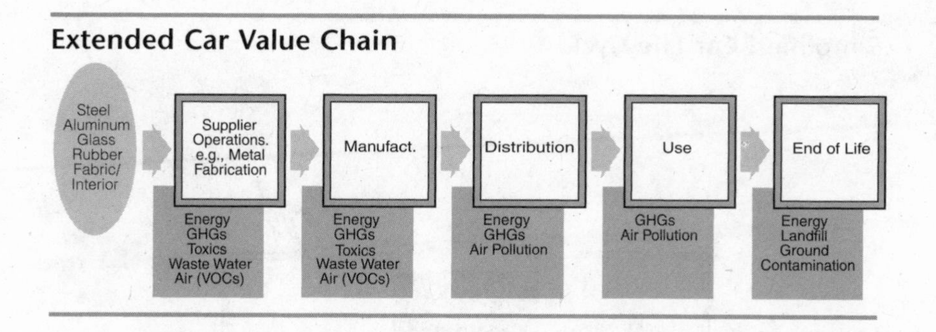

Some aspects of a company's footprint are positive, such as customers served, employees paid, and support provided to communities. But in the environmental context, the footprint reflects how much a company's pollution burdens society and the scale of its consumption of the Earth's natural resources. Increasingly, society is insisting that those who tax the planet's capacity to absorb human impact both pay for the damage they do and work to minimize the harm. WaveRiders take the measure of their footprint seriously and set explicit goals to reduce its negative effects.

But before they can reduce their footprint, they need to know its contours. One of the most useful tools for measuring the footprint is the product Life Cycle Assessment, or LCA.

Life Cycle Assessment (LCA)

Life Cycle Assessment tracks the environmental impacts of a product from its raw materials through disposal at the end of its useful life. LCA is an important tool for developing an environmental self-portrait and for finding ways to minimize harm. A good LCA can illuminate ways to reduce the resources consumed and lower costs all along the value chain.

Let's take one more look at that car value chain (see figure).

Instead of a linear path, we now see a circle. The car starts as raw natural resources extracted from the earth—everything from iron ore and aluminum to cattle for leather interiors. Suppliers receive these inputs and craft the various car parts. The car company assembles its product and distributes it globally. After 200,000 miles on the road, the car enters the end-of-life phase. Some pieces will be sold

Simplified Car Life Cycle

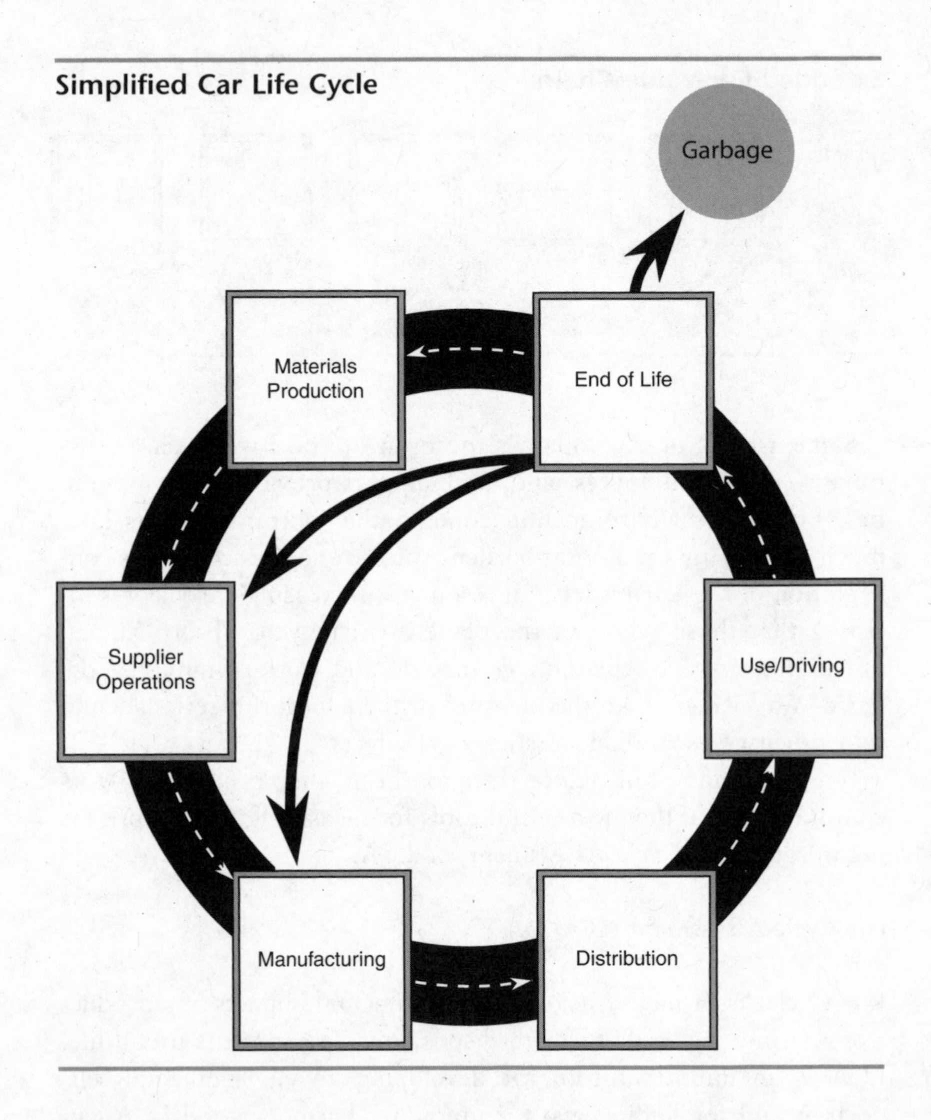

for scrap back to raw material producers, other parts refurbished and reused in another car, and still others sent to the landfill.

A Life Cycle Assessment looks at this complete circle and measures environmental impact at every phase. It provides the foundation for understanding the issues a company must address and clues to help find Eco-Advantage. Which step in the process uses the most water, or produces the most air pollution? Can we reuse or recycle any by-products from manufacturing? Can we recycle the whole product? Which steps in the chain create environmental impacts that would concern different stakeholders? Where is waste and inefficiency? The answers can be surprising and vary widely across products and industries.

Take, for example, one politically charged environmental impact, greenhouse gas emissions. Imagine we have only three life cycle stages: pre-manufacturing (before you), manufacturing (you), and use (after you). To oversimplify, that's suppliers, your operations, and customers.

Now let's look at the approximate greenhouse gas emissions for three products: an SUV, a bank account, and a leather boot (see figure). The greenhouse gas emissions profiles for these three products are vastly different. For the SUV, the majority of the greenhouse gases come not from manufacturing, but when drivers burn fuel (the use phase). For a service business like the bank branch where the account is held, energy use in operations (lighting, heating and cooling) is the biggest percentage of its own immediate footprint. And for the boot, most of the emissions occur upstream in the supply chain.

The numbers here are not exact since every product or company is different, but they are very much grounded in reality. In fact, Timberland conducted an LCA for one product, a leather boot, and calculated emissions at all stages in the value chain, including subsuppliers like the cattle industry (again, they provide the leather). Surprisingly, the cattle are responsible for the most emissions by far. During digestion, cows produce methane, a powerful greenhouse gas.

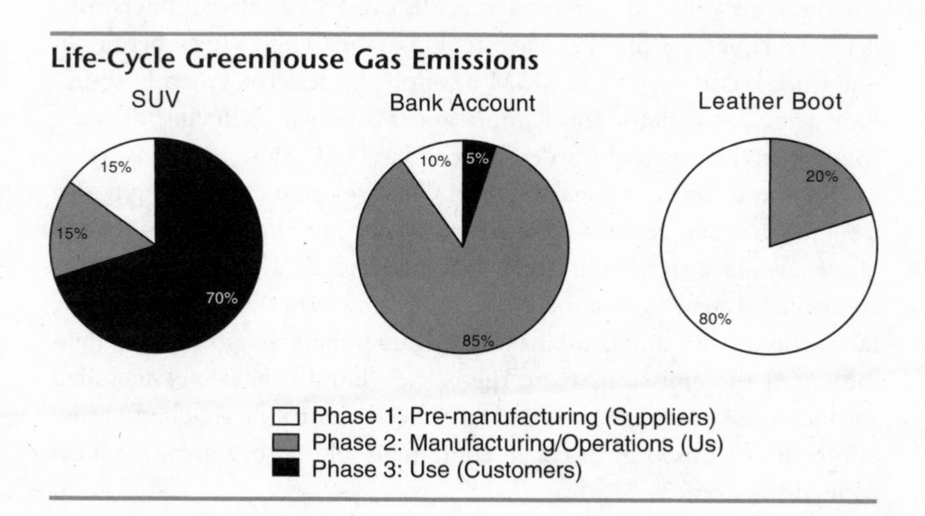

Cow hides are only a fraction of the weight of the cow, so Timberland assigned only seven percent of the cows' gas production to the boot's environmental balance sheet. Yet even at that small percentage, cattle are responsible for 80 percent of the greenhouse gas burden of the boot.

Why does this matter? In theory, armed with better information about the size and nature of the various environmental aspects of its boots, Timberland could focus its efforts in the right places. Or at least executives are better positioned to understand the trade-offs they face. Nobody in the company is thinking seriously about removing all leather from its boots, and nobody can stop a cow from churning out methane. But Timberland now knows that reducing the amount of leather per boot will shrink its climate-change footprint far more than reducing energy use at assembly plants or distribution centers.

LCAs can also guide product development. 3M has been rapidly expanding its Life Cycle Management program. The protocol for new products now includes an assessment of environmental, health, and safety issues at suppliers, within 3M, and with the customer. This understanding of the full systemic impacts of its products gives 3M a powerful foundation on which to build strategies for Eco-Advantage. Take 3M's favorite tool, eco-efficiency. If you know the pressure points in the system, and you know where you can have the most impact, your strategies get that much better. You can cut pollution and waste with a scalpel rather than a chain saw.

Companies that understand the life cycle of their products also drive revenues by finding ways to make customers' lives better. In one case, based on an LCA, 3M developed a less toxic, ready-to-use industrial disinfectant for schools and hospitals. Reducing the customers' environmental burden drove increased sales.

One important cautionary note: LCAs are often challenging to do. The analysis can cover a product, a division, or the whole company. Some calculations are relatively straightforward, while others require careful analysis—and some big assumptions. One core issue is how far upstream or downstream in the value chain to go. Timberland measured the emissions from the cattle. Should they have included the emissions from the fuel used to deliver feed to the ranchers? How about the fuel used in the farm equipment where they grew the corn to feed the cattle?

Consider the trade-off between depth and cost. Simple analyses often yield significant insights. So remember the 80/20 rule—80 percent of the work comes from 20 percent of the issues—before plunging into a deep LCA. If you don't make reasonable assumptions, draw the boundaries in a logical way, and acknowledge the limits of the analysis, LCAs can very quickly spin out of control. Some amusingly thorough LCAs have included the energy needed to produce the food eaten by people involved in making a product.

Ultimately, knowledge is power. Understanding its footprint along the value chain can help a company engage in pollution prevention, find new opportunities to serve customers, avoid problems with stakeholders, and gain a big leg up on the competition. That's Eco-Advantage.

WHERE ARE WE HEADED?

One of the biggest challenges in the sustainability realm is simply defining what it requires of corporations. One organization, the Natural Step, has a useful framework with four principles that help companies create a picture of what it means to be sustainable. (See endnotes for the Natural Step's "system conditions.") Working with a clear vision of the endpoint, companies then can begin to lay out the steps necessary to get there—what Natural Step founder Karl-Henrik Robert calls "backcasting from principles."

A number of WaveRiders—McDonald's, Starbucks, and Interface, to name a few—credit Robert's approach with helping to advance their thinking about sustainability. IKEA executives, for example, pictured a specific future where lighting was low cost and environmentally safe. Current low-energy lamps were expensive and contained toxic mercury, which violated Robert's system conditions. So the IKEA team went searching for new options, resulting in a refined production process that cut mercury by 75 percent. Asking themselves "Where are we headed?" focused attention and drove innovation on a long-standing problem.

CAPTURE DATA AND CREATE METRICS

"What gets measured gets managed." Maybe that claim is too universal, but metrics—particularly when tied to rewards—attract attention. To manage any aspect of business performance you need information. In the environmental world, data has not always been at the center of decisionmaking, strategy, or policy. It should be. Data, after all, is almost always an important precursor of real, verifiable environmental improvement.

Just one data tracking law—the Toxics Release Inventory program—started many WaveRiders down the path to environmental leadership. After being required to measure and report their emissions of a long list of toxic substances, companies saw just how big their footprints were. And many realized that valuable chemicals were going up the smokestack.

What to Track

Every company should track some basic environmental outcomes—the resources it uses, and what it emits or wastes. Leading companies track their performance in a number of areas (see table).

It would be nice to distill the outcomes and get to a single metric, but it's not possible. Herman Miller's CEO expressed frustration to us that he couldn't get environmental metrics down to a couple of key items for his "balanced scorecard." But the fact is, protecting the environment is inescapably a multidimensional challenge.

Environmental indicators are like financial metrics. Everyone produces the basic tools of financial measurement: the income, balance sheet, and cash flow statements. But each company picks specific metrics to focus on, such as net income, debt-to-equity level, or free cash flow. We need all of these measures, but some are more essential than others, depending on the circumstances.

For those using a balanced scorecard to manage their business, our list of metrics provides a good starting point for what should appear on the environmental side of the ledger. Those looking at triple bottom line performance will have to include social metrics as well.

Every company should tailor its data collection to fit its own core issues and develop company- or industry-specific indicators. Starbucks, for example, closely tracks several aspects of its paper usage, such as percentage recycled or unbleached content. Millions of paper cups add up. To keep an eye on an issue that could squeeze operations, Coca-Cola tracks how many liters of water it uses to produce a liter of final product.

Key Environmental Metrics

Environmental Outcome	Basic Metrics
Energy	— Energy used — Renewable energy used or bought
Water	— Total water used — Water pollution
Air	— Greenhouse gas emissions — Releases of heavy metals and toxic chemicals — Emissions of particulates, VOCs, SOx, and NOx
Waste	— Hazardous waste — Solid waste — Recycled materials
Compliance	— Notices of violations — Fines or penalties paid

We suggest three general guidelines for environmental data and metrics:

- **Track both relative and absolute metrics.** Measure greenhouse gases, for example, per dollar of sales and total tons of carbon dioxide emitted. It's tempting to show progress in relative terms, but some environmental problems are absolute. Reducing greenhouse gas emissions relative to sales is useful. But if sales increase significantly, the problem is still getting worse. Key stakeholders won't be impressed by a denominator-driven claim of progress.
- **Capture data at multiple levels within a company.** The ability to drill down—by country, region, division, site, and even production line—can help isolate problem areas to address or highlight leading-edge performance and best practices to replicate.
- **Collect the same information for the whole value chain.** At the risk of sounding like a broken record, what happens outside the factory gates, from suppliers to distributors to customers, can be critical.

If all of this sounds expensive, it doesn't have to be. Good measurement often goes hand in hand with tight operational control. Even tiny Rohner Textil, one of the smallest companies we interviewed, has tracked dozens of environmental metrics since 1993. In a moment, we'll see how GE has saved money with an elaborate eco-data tracking system that offers a high degree of granularity.

Supply Chain Metrics and Materials Databases

So far, we've talked about metrics best suited for businesses with concrete products. A service business uses energy and may generate waste, but probably doesn't need to track water emissions. But even service businesses rely on products, and their suppliers might face some different issues. Tracking supplier performance is a good idea. Could a bank be held liable for environmental issues created by the manufacturer of ATM machines? It may sound ridiculous, but who would the NGOs and media go after—the obscure manufacturer or the big customer-facing multibillion-dollar brand?

McDonald's has starting asking its suppliers to track key metrics. The company's own operational footprint is wrapped up in energy use and waste. But burgers and fries come with environmental baggage. Cattle raising and industrial agriculture have large footprints—a point brought home to McDonald's by the many questions posed by NGOs about whether its meat came from grazing lands that were cleared from rainforests. McDonald's realized it needed to get a handle on those impacts as well.

Some WaveRiders are going beyond basic metrics and asking suppliers for proprietary information. Herman Miller built a materials database to guide its production of the environmentally sound Mirra chair. The two managers leading the charge asked every supplier to give them the exact ingredients in every component. Some companies balked at sharing this information, but Herman Miller would buy only from those suppliers who complied.

Every single chemical and substance received a score of red, yellow, or green based on toxicity and other environmental attributes. Management told designers that they could use greens with no qualms but had to try to minimize yellows and avoid reds like PVC plastic. With this database of 800 materials—who knew so many things went into one chair?—Herman Miller can calculate with precision the quantities of every substance in every chair. Since the tool also generates an overall product score from 1 to 100, designers now have targets for each new generation of products.

Data management is a critical tool for generating Eco-Advantage in all categories: risk, cost, revenue, and intangibles. Customers looking for furniture with low environmental impact and no concerns about indoor air quality will be satisfied by the Mirra. Cleaner products enhance the brand as well. Finally, knowing precisely what's in your products can slash risk and save millions of dollars.

Sony's cadmium crisis was a wake-up call heard broadly across the electronics industry. Dell, for example, is spending millions to beef up its materials database. One of Dell's key environmental executives, Don Brown, told us, "If there's a problem when we hit the dock in the EU, we can answer any questions with *data* and avoid holding up 10,000 units in customs."

Going Past Raw Data to Creative Metrics

Having good data gets you to the starting line. Using the information to demonstrate environmental impact in interesting, relevant ways can focus employee attention on the right things. Talking about energy use per employee, for example, brings the challenge down to the individual level and grabs everyone more than a grand total.

DuPont has focused on a deceptively blunt metric: Shareholder Value Added (SVA) per pound of product. That's right, the weight of everything the company makes, which, they say, was once in the vicinity of 20 *billion* pounds. DuPont is getting at a profound truth: even with aggressive goals on waste and energy, the more stuff it produces, the bigger the environmental impact. So the company cut to the chase and just measured the volume. DuPont has six years of data on SVA per pound for ninety divisions. The informal goal is to quadruple this measure of resource productivity company-wide.

In 2004, DuPont lost some of its iconic brands including Lycra and Spandex when it sold its nylon business to Koch Industries. How does this sale relate to pounds of stuff? DuPont's Paul Tebo explains it in market valuation terms. He draws a theoretical chart of SVA/pound against the Price/Earnings ratio of different companies or industries. At one corner is a company like Microsoft—high P/E, high SVA/lb. At the other end are big, heavy industries like nylon production—low P/E, low SVA/lb. DuPont believes the two axes are connected and wants to move to the high value-added end of the spectrum, with higher growth rates and higher valuations.

We will always need products and materials, but the markets value more highly the services and knowledge added on top of these materials. As we shift to an information-based economy, bytes replace bits. So, in a way, SVA/pound captures the spirit of the broad global move toward dematerialization. That's a lot to heap on one little metric, but it does the job.

Data and Competitive Spirit

Rankings and lists start arguments. Just look at the hullabaloo over *U.S. News*'s annual ranking of colleges. Or the yearly flap over the coaches' poll and computer ranking that determine which college football teams go to the major bowl games. As we've found out, even dry environmental data can raise a big fuss.

Our research team at Yale, working with the Earth Institute at Columbia University, created a ranking of countries based on sustainability metrics. The results, released at the annual meeting of the World Economic Forum, touched off a firestorm in a number of countries. Then Mexican President Ernesto Zedillo insisted that his Environment Ministry officials visit Yale to complain. The Belgian Prime Minister faced a parliamentary inquiry, and Singapore's Environment Ministry suggested that it was impossible, perhaps even unconstitutional, for Singapore to be ranked near the bottom.

WaveRiders know that data can stoke competitive fires. They compare environmental performance across sites, regions, or divisions. Latin American conglomerate GrupoNueva releases all its metrics internally, every month, by facility. Plant managers know where they stand on a wide range of issues including environmental performance. IKEA's internal reviews of store managers include their rank on environmental metrics within their country and continent. It's just human nature: Nobody wants to be at the bottom.

Environmental metrics show a company where it stands. Data and indicators are critical to fact-based decision-making and sound environmental management. They drive continuous improvement and allow managers to mark progress against pollution control and resource productivity goals. Sustainability is more a journey than a destination, but it still pays to know where you are on the path.

SET UP SYSTEMS

Tax time always brings out two basic types of people: those who have all their receipts, bank and brokerage statements, and mortgage stubs sorted and ready to go . . . and the rest of us. The information exists in old shoe boxes and at the bottom of desk drawers, but we've set up no precise system to capture it.

Businesses don't have that option. Financial statements must conform to Generally Accepted Accounting Principles, and public companies are required to report on their performance. In the United States under the Sarbanes-Oxley Act, CEOs and CFOs could face jail time if their financial statements are not in order. Environmental issues are now considered part of the potential liabilities that companies must include in any such accounting.

Implementing an environmental management system (EMS) is thus essential for companies of all sizes. Setting one up takes time and care, but when a good system is in place, managers know their business better, find ways to squeeze out waste and make processes run more efficiently, and avoid serious problems. Take the example of GE. It's no surprise that the company, with its data-oriented corporate culture, has built an elaborate system for tracking environmental results. Senior Vice President Steve Ramsey walked us through GE's

PowerSuite, an intranet program that provides detailed process information, like a regulatory calendar that reminds managers what to do and when to file to meet permit requirements.

The real-time "digital cockpit" includes metrics on environmental performance, resource use, safety, and compliance at almost every level—from company wide to the granular level of a discrete production process. If there's a hitch anywhere along the line, plant managers and their bosses will know and can act on it.

GE's tracking system cost about $10 million to develop. But this data emphasis, linked with the company's famed commitment to Six Sigma efficiency, has produced outstanding environmental management results. GE has reduced the number of times wastewater emissions have exceeded levels permitted by regulations by more than 80 percent in a decade. The company has also saved tens of millions of dollars through environmental and safety productivity improvements. The new system quickly paid for itself many times over.

Environmental Management Systems

Setting up an environmental management system doesn't have to be a solo journey. Standardized EMS platforms have already been developed. One particularly useful place to look for help is the International Organization for Standardization's environmental management standard, ISO 14000. Like the well-known ISO 9000, which guides companies in building quality into their operations, ISO 14000 provides a template for setting up an environmental management system. Not every company with good data and systems needs ISO 14000, but many now seek certification at the "request" of their big customers.

Help in implementing ISO 14000 can be obtained from a variety of sources. Both the World Business Council for Sustainable Development and the Global Environmental Management Initiative provide guidance. And alternatives to ISO 14000 are available. Some European facilities have adopted the European Union's Eco-Management and Audit Scheme. A number of the companies we studied have developed their own systems, often with more emphasis on improving outcomes than the process-focused ISO requires.

So while creating a detailed management system may seem unexciting or expensive, our research shows that having *some* system in place is well worth it and drives optimal performance.

When Things Go Wrong

Remember how badly Exxon executives floundered in the days after the *Exxon Valdez* oil spill? The company ended up spending $3 billion to clean up the oil in Prince William Sound. Additional penalties, fines, and legal judgments so far have sent the total tab over $12 billion. And the books on this 1989 incident have not yet been closed.

One last specific systems area deserves special focus: risk assessment and emergency management. Say something goes wrong at one of your facilities—a spill of something toxic or, worse yet, actual injury or death. What should the company do? Who's in charge of fixing the problem? Who talks to the press? Who talks to employees? These questions should *not* be answered on the fly. In a crunch, time is of the essence. When accidents occur, companies need a clear set of procedures. Everyone involved in an emergency response must know his or her role. IBM, for example, has a corporate crisis management team with responsibilities spelled out and specific managers defined as acceptable spokespeople.

Well ahead of a crisis, companies need a process for identifying potential environmental risks. The AUDIO tool we described briefly in Chapter 2 is a good start. It helps you figure out where in the value chain crises might erupt.

A number of tools can help sharpen thinking and focus, particularly at the day-to-day operational level. We've worked with companies like Northeast Utilities to implement simple questionnaires on environmental risk. The questions push the critical middle managers to think about where the company's challenges and vulnerabilities are. Do they know how they'd handle emergencies? Is the necessary equipment in place to respond to an accident? Have personnel been trained and drilled? The questionnaire helps generate a gap analysis to identify holes in the system.

WaveRiders think about environmental risks before they become problems. IKEA's Material Risk Council, for example, meets regularly to review the latest thinking on chemicals. Those deemed too dangerous are put on a blacklist and eliminated from the supply chain.

Companies face an especially critical point of exposure when they acquire new assets. With a substantial volume of mergers and acquisitions, GE conducts a rigorous environmental and risk assessment on all deals. The company builds into the price of the potential new asset the costs for (a) bringing the new business into full regulatory compliance, (b) putting GE's Environment, Health, and Safety systems into place, and (c) dealing with legacy liability issues. Key members of the environmental review team meet monthly with management to discuss deals in the pipeline.

The risks identified can kill a deal or result in mandatory remedial action. The GE team tells a chilling story about a Brazilian firm the company bought a few years ago. During the deal review, the team noted that the on-site day care was near an area with substantial chemical storage. GE demanded that the company move the kids. About three months later a chemical fire broke out and destroyed that part of the building. GE executives have no doubt that many children would have been killed. "This example is why you need to be systematic," the GE team told us. A finance-only review of the deal would not have uncovered the threat.

PARTNER FOR ADVANTAGE

You might think you have a handle on your big environmental issues if you've set up tracking tools and systems. But NGOs and local communities often have a very different view of your issue set. Remember, feelings are facts. We can think of no better way to track what outsiders think of your company on an *emotional* level than by partnering with, and really listening to, outside organizations.

Partnering is, in fact, a key tool for generating Eco-Advantage. Whole books have been written about partnerships, and many conferences have been devoted to the topic. Here, we give a brief review of notable examples and provide key lessons we've gleaned from analyzing dozens of business–stakeholder relationships.

In theory, a company could partner or learn from nearly any of the twenty stakeholders on our Eco-Advantage Players wheel from Chapter 3. It's worth thinking through the possible links with all of them. But five key categories of partnership are most likely to pay big dividends: NGOs, environmental experts, governments, communities, and other companies.

Partnering with NGOs

If there were a contest for the company with the worst corporate social responsibility history, Chiquita (formerly United Fruit), would definitely be in the running. In the 1950s, the company funded a CIA-led coup of the democratically elected government in Guatemala because it disliked the new leaders' agriculture policies. (Hence, the phrase "banana republic.") This tortured history makes the turnaround of Chiquita, now a leader in environmental and social responsibility, all the more remarkable.

Chiquita executives will tell you that the transformation began in the early 1990s when they formed a deep, lasting, and unlikely partnership with the Rainforest Alliance, a New York–based NGO. The key players in this relationship are still there today. We spoke with Dave McLaughlin from Chiquita and Chris Wille from Rainforest Alliance, both of whom work in Costa Rica and spend significant time on the farms that produce the world's bananas.

Today, the partnership is multifaceted and deeply entrenched. Working with a range of other stakeholders, including some other companies and small farmers, Chiquita and Rainforest Alliance designed a set of guidelines for growing and processing bananas in an environmentally and socially preferable way. After a two-year intensive and inclusive process, they produced a new way of doing business. Now, every year, Rainforest Alliance audits and certifies farms that meet the standards of its Better Banana Program. Chiquita then shares the results with the public in a brutally honest environmental report, including notes on even the smallest failures. One example: "In Turbo, Colombia, a company automotive facility was leaking oil, which could have entered a nearby stream."

This astonishing transparency has won awards for environmental reporting excellence and earned Chiquita loyalty from watchful customers in Europe. The reports were years ahead of the more recent efforts from other companies like The Gap, which has also opened its books on supplier performance.

The Chiquita–Rainforest Alliance partnership is one of the most strategic and effective in the world—and one of the oddest, at least from the outside. As the 1990s began, Rainforest Alliance was among the biggest critics of the banana industry generally and Chiquita in particular. If Chiquita were the Roman Empire, executives saw the NGO as the raiding hordes of barbarians.

NOTABLE QUOTE: STRANGE BEDFELLOWS

As Conservation International's Glenn Prickett notes:

> Lines between these camps were once clearly drawn. Now former antagonists work together in ways that are uncomfortable, controversial, and yet often highly effective. . . . In their heart of hearts, most environmental NGO leaders would probably prefer public policy solutions to industry partnerships. And most industry executives would probably prefer to focus on business, not environmental work. But we live in an era of strange bedfellows.

In 1991, the Rainforest Alliance produced a preliminary set of standards for how farmers should grow bananas. The industry rejected it. As McLaughlin told us, Chiquita thought it might be a trap. Executives believed that the increased visibility of the company might make it a target of criticism, despite their best intentions. Leaders at the NGO acknowledge that this fear was not unfounded. "Many NGO criticisms were unfair or generalized," Chris Wille admits. "All bad things happening in Costa Rica were laid at the feet of one industry and one company."

Relations, in short, were virtually nonexistent. But many forces were conspiring to bring Chiquita to the table. European buyers were asking tough questions about banana farming, and Chiquita knew it needed to make a good faith effort to improve. Working with the Rainforest Alliance offered a credible starting point. The relationship started slowly with a simple meeting. McLaughlin and Wille walked a farm together to explore issues. After a gradual building of trust, Chiquita agreed to develop a two-farm pilot program to test new standards and learn what could really work on the ground. But the relationship was still dicey, even at times icy. As we mentioned earlier, *Ranger Rick* magazine led a kids' write-in campaign to the Chiquita CEO. The Rainforest Alliance was behind the effort, and this was *after* they had started working together.

But trust continued to grow. Over the years, the partnership has combined sound science with on-the-ground realities and operational savvy. The Better Banana Program is a detailed operational playbook. Chiquita spent $20 million over the first decade to roll out changes across the continent, but saved $100 million in operating costs at the same time. Farm productivity is up 27 percent, and cost per box of bananas is down 12 percent. Chiquita managers have no doubt that these farms just plain run better.

The environmental results are impressive as well. Chiquita farms have drastically cut the use of pesticides, eliminating some insecticides entirely. The farms now carefully manage waste such as plastic and bad bananas that they once dumped haphazardly. "It's night and day on the farms," Wille tells us. "The plastic litter used to be knee deep and wash into rivers. Tourists who know nothing about the program say they can spot a certified farm because it is cleaner and better organized and managed."

The benefits in terms of employee morale are enormous. Banana industry executives have told Wille that the difference in worker attitude alone has made the expense of the program worthwhile.

NOTABLE QUOTE: CHANGES AT CHIQUITA

All of the operational changes on Chiquita's banana farms are important. But Chiquita executives really rave about the changes in their own company. As Dave McLaughlin says,

> At the end of the day, it's not the certification that interests me. What is truly the most valuable part is the process. Getting your people plugged into it and thinking outside the box and getting new ideas . . . all of those things are ten times more important. To see our senior managers become friends with the NGO is something. Our management is far more enriched than they were. In our first CSR report, a letter from the CEO talked about how engagement with the Rainforest Alliance led us down a completely new path. We opened that door and danced with the devil. And we're better off for it.

The other gold star for NGO partnership goes to McDonald's. For over fifteen years, the company has worked with a who's who of global environmental NGOs. First came the groundbreaking partnership with Environmental Defense on packaging. Just one change, the move away from the Styrofoam burger "clamshell," was international news. More recently, McDonald's has worked closely with Conservation International to examine the environmental impacts of its supply chain.

Together the two organizations established a set of principles for a pilot program, including a systems-based approach, a long-term view, and a base in real science. With these familiar elements of the Eco-Advantage Mindset, they hoped to understand McDonald's upstream impacts on water pollution, soil erosion, and waste management, especially from livestock. The partnership brought in some pilot suppliers to test new metrics and goals. We spoke with the company's largest beef and bacon supplier, who found the pilot program tough but enriching.

Cynics might contend that McDonald's is simply co-opting the agenda of these NGOs. We say, so what? It's not some Machiavellian conspiracy if a company takes on environmentally sound practices that NGOs help them design. McDonald's is working in good faith with these organizations. It's true, public criticism of companies with whom NGOs partner is way down. But then again, there is less to complain about. And it's much harder for the NGOs to attack their own partnerships and programs. Partnering gives a company a strong defense against NGO attacks, but a large part of that defense is the demonstration of genuine progress. We call it **brand inoculation**—another form of Eco-Advantage.

As true partners, NGOs can help companies track what's coming over the horizon and how the brand is perceived. Picking their brains and really listening will give a company an Eco-Advantage over competitors who ignore what this important part of the market is saying.

Partnering with Experts

WaveRiders also seek Eco-Advantage by partnering with knowledge generators. Academics and other environmental experts can provide valuable perspectives on growing issues of concern. Launching a dialogue, or even opening operations to their scrutiny, provides a peer review mechanism and a way to keep the company at the cutting edge.

Alcan cautiously accepted the request of some university researchers to study its (now former) aluminum operations in Jamaica. "They crawled all over our operations," Alcan's Dan Gagnier, told us. "Then I held my breath for the report." Looking at the full life cycle, the bauxite operations were declared "best in class," which was a relief. But not everything was perfect. The report helped Alcan improve on issues like rehabilitating the land after mining. Alcan let experts in and got an in-depth LCA in return for its transparency.

DuPont regularly invites experts to judge its internal Sustainable Growth Excellence Awards. The awards celebrate employees and projects that moved the ball on sustainability. But DuPont uses the review process as a scouting exercise. They bring in a range of scholars, researchers, and NGOs, including some people who would normally be antagonists. The awards panelists ask tough questions and let the company know if they are off track.

Other companies do their scouting more directly. Dow, Unilever, Coca-Cola, and a growing number of corporations, big and small, have created environmental or sustainability advisory boards to meet regularly with company officials. This peer review gives companies feedback on their environmental programs and performance from an independent perspective. And they get a chance to hear from top NGO leaders, academics, and environmental management experts about emerging issues and environmental community priorities. This scan of the horizon provides a critical heads-up on what might be coming down the pike.

Finally, a very few WaveRiders have gone even a step further and brought environmental science expertise in house. Clif Bar hired an ecologist trained at Yale to work full-time on the company's sustainability efforts. She has worked with every department, helping the company move to organic ingredients and focus more of its marketing efforts on global warming issues. The company has shifted some of its sponsorship of sporting events to focus on environmental issues, for example, by making a bike race "climate neutral." This internal perspective from a trained scientist has helped the company find new ways to connect with health-focused, outdoorsy customers.

Partnering with the Government

For many years, the government's role in environmental issues was clear. Congress passed the laws, and regulatory agencies promulgated implementing rules. "Command and control" regulation hasn't disappeared. But today both the federal EPA and state departments of environmental protection often look for ways to work collaboratively with business. The EPA has launched dozens of voluntary programs and industry partnerships. Some initiatives such as the Energy Star ratings on computers and appliances set standards for using a government-approved label. Others have more of a give and take. (A more complete list of government programs is on our website, eco-advantage.com.)

Intel was one of the first companies to take the EPA up on its Project XL offer of regulatory flexibility for companies willing to show "excellence and leadership" and commit to pollution controls above and beyond legal requirements. With the advice of a stakeholder group, Intel established a set of stringent environmental goals, reviewed them regularly, and released quarterly metrics. In return, Intel received a site-wide permit and top-speed regulatory review for its expansion plans. Intel executives call this initiative a "huge win."

With the European Union taking an aggressive stance on advancing environmental regulation, partnering in that region might prove particularly valuable. Nokia, for example, is one of two companies (the other being Carrefour) leading a pilot project with EU officials under the new Integrated Product Policy. The IPP focuses on the lifecycle impact of products. The pilot program is bringing together a range of stakeholders from Nokia's competitors to recyclers and customers to explore how to reduce the total environmental impact of cell phones.

Partnering with Communities

Communities are no longer bystanders when it comes to corporate actions. Every day in cities, suburbs, and rural areas across America, would-be expansions and new facilities fall victim to antisprawl campaigns and "not in my backyard" attitudes.

Shell, like other resource-intensive industries, used to make deals with governments or land owners and move forward with major projects. No longer. As we mentioned earlier, Shell's work with the community near its Athabasca oil sands project in Alberta, Canada, offers a new model for how things must be done.

The Tar Sands are the great oily hope of fossil fuel lovers. Estimates suggest the possibility of oil reserves nearly as big as Saudi Arabia's. But the process of extracting the oil represents a major environmental challenge. It's *everything*-intensive—people, land, energy, water, money, time.

Reasonably enough, local native communities have questions. What's going to happen to our air? What will happen to the water table and local rivers and streams? Fort McMurray, the population center, has already grown from a hamlet of a thousand people in the 1960s to a bustling city of 35,000 in the 1990s. It's projected to shoot past 100,000 residents if the oil sands project goes forward, with all the additional environmental pressures such rapid growth brings.

To allay the fears and prepare for the future, Shell worked closely with the native communities, helping them build businesses to serve the project, preserving burial grounds and fishing rights, and even teaching them to prepare résumés so that they could take advantage of the job boom. The company created regional infrastructure plans with NGOs, communities, and competitors. "We did things not required by law, or even by Shell internal HSE practice at that point," Shell's Mark Weintraub told us. "All this work has helped position us for approvals and expansion as we become a 'partner of choice' in the region." Shell has spent a lot of time and money, but Weintraub believes that the approval process was shortened by at least a year, a substantial saving in an industry where time really is money.

Partnering with Other Companies

Some environmental issues can't be addressed by a single company on its own. Sometimes, it makes sense for an entire industry to take on a problem or for a group of companies with similar interests to work together on solutions. Cross-company partnerships take many forms:

- **Agreement on shared learning.** McDonald's, Coca-Cola, and Unilever created the Refrigerants, Naturally! initiative—supported by the United Nations Environment Program and Greenpeace—to explore alternatives to chemicals in refrigerators and freezers that deplete the ozone layer and contribute to global warming. The U.S. EPA recognized the initiative with a Climate Change Award in 2005.
- **Closing loops.** Albertsons supermarkets won an America's Marketplace Recycles Award for its innovative supply chain partnerships focused on reuse of wooden pallets, cooking oil, and other materials.
- **Commitment to change a market.** Microsoft, Kaiser Permanente, and Crabtree & Evelyn worked with a group of sixty companies to phase out PVC plastic from products and packaging.

Partnerships with big ambitions, such as remaking a market, require real effort and somebody willing to step up and take the lead. Working with the NGO Metafore, a diverse list of companies—including Staples, Time Inc., Starbucks, FedEx Kinko's, Nike, and Toyota—have come together to form the Paper Working Group. After struggling on their own to define environmentally responsible paper and find suppliers that can provide it, the companies came together to harmonize the tools they use to evaluate paper purchases. Their combined buying power—together they represent more than one percent of global paper demand—should also increase the supply and affordability of green paper.

In 2004, leading electronics producers, including HP, Dell, and IBM, launched the Electronics Industry Code of Conduct to harmonize supplier standards for the entire industry. A joint code of conduct produces scale economies in implementation and training, which saves suppliers lots of time and money. Other industries have noticed

WORLD BUSINESS COUNCIL FOR SUSTAINABLE DEVELOPMENT

For nearly fifteen years, the WBCSD has been developing company and industry partnerships. As executive director, Björn Stigson has helped companies in a range of industries band together to share best practices, particularly in the realm of eco-efficiency, and to develop strategies for sustainable development.

what the electronics companies are doing. From pharma to automakers, companies are quietly putting together joint policies for managing their supply chains.

In some contexts, one company raising its standards alone might be disadvantaged. Partnerships make perfect sense. The burden of being out in front—and bearing environmental costs others are ducking—will be especially acute in highly competitive industries where even slight cost differentials can kill a business. In these circumstances, smart companies look to establish industry-wide partnerships. Leading-edge companies that are prepared to meet more demanding requirements might also find that quietly lobbying the government for more stringent regulations is the best path forward.

Since industry partnerships eliminate the opportunity to seize a competitive edge, cooperation will be strategically optimal where a company seeks to level the playing field or where acting collectively saves everyone money and pushes the industry forward. Partnerships don't make sense when solo action offers a basis for seizing Eco-Advantage.

Eight Lessons Learned on Partnering

In analyzing dozens of partnerships, some successful—and some less so—we've identified a series of fundamental lessons:

1. KNOW YOUR OWN SITUATION WELL BEFORE PICKING AN APPROPRIATE PARTNER

Be clear on your environmental issues when you sit down with others. An AUDIO analysis is a good place to start. Then educate yourself on your business's key problems, and learn which groups specialize in the issues you face.

2. KNOW WITH WHOM YOU'RE DEALING

All partners, especially NGOs, are not created equal. Sustainability expert John Elkington has developed a playful, but useful, typology of NGOs. He breaks them into sharks, orcas, sea lions, and dolphins. Sharks are always on the attack, smelling blood and weakness from miles away. Orcas use fear and bullying. Sea lions play it safe and stay close to issues they know well. Dolphins are intelligent, creative, and can help fend off sharks. The point is that some NGOs are easier to work with than others. Avoid the sharks.

3. BE PATIENT

If we could share only one lesson, this would be it. Trust builds over time. It can take years to make the case internally for reaching out. As Chiquita's Dave McLaughlin says, "We aren't making Tang here. It isn't just 'add water and stir.' " Nurture long-term relationships.

4. LEARN EACH OTHER'S CULTURE AND VALUES

IKEA spent six months with World Wildlife Fund just discussing values before launching a partnership. The differences between for-profit and not-for-profit organizations can be large, but different values and cultures are not insurmountable. Still, it takes effort to learn to talk the other guy's language.

5. SET WORKABLE GOALS

Partnership goals need to be carefully developed and specified. They must achieve environmental progress that satisfies all the partners, but also be relevant to and supportive of core business objectives. Set modest short-term goals and exceed them. And never over-promise publicly.

6. ESTABLISH CHAMPIONS

Each partner needs a clear operational leader for the project and the relationship. Backing from the highest level is also vital. IKEA reports regularly to the CEO on its World Wildlife Fund partnership. You also need critical line managers to climb on board. McDonald's work on its supply chain took off only when the supply chain managers, not just the corporate responsibility people, stepped into the process.

7. THINK BIG, BUT START SMALL

The commitment to green the supply chain is a worthy goal, but it can't be done overnight. Pilot programs provide a way to test assumptions, establish trust, and build a base for bigger and broader future partnership initiatives.

8. COORDINATE COMMUNICATIONS

Great partnerships can turn sour very quickly when one side prematurely declares victory. NGOs see greenwashing and companies hear gloating. You can't assume that the way you would talk about an issue is how the other side would. In the same spirit, don't an-

LESSONS FROM THE CORPORATE TRENCHES FOR NGOs

A progressive NGO will often meet its goals and help clean up the environment more quickly if it puts down the sword and finds a way to work productively with companies. We've identified a few guidelines for greater success:

- **Don't just criticize.** Companies listen better if they hear encouraging words.
- **Be constructive.** No company is sustainable, so it's easy to tear them down. But can you build them up?
- **Don't expect success if you equate all business with evil.** Anti-market or anti-capitalist approaches don't provide a good foundation for corporate engagement.
- **Understand the pressures business faces.** Give company officials cover by building a credible, practical plan that makes them look good with their management.
- **Map *your* stakeholders.** And watch your backside. NGOs can be accused of selling out when they work with companies. While building the banana program, Rainforest Alliance found itself getting more support from industry than sister NGOs.
- **Let some things go.** Compromise is essential to a working partnership. Remember that "the perfect is the enemy of the good."
- **Admit mistakes.** And don't be afraid to revise views or positions, even long-standing ones.

nounce environmental breakthroughs until you have credible evidence of progress.

Overall, if there's one clear message to take away from partnerships, it's this: Do not be afraid to start a conversation with your harshest critics. Listening to an extensive critique may be painful, but it's better than finding it splattered all over the web. At the very least, all the parties will learn something. Knowing where you stand with everyone is a key component of eco-tracking and the first step in building competitive Eco-Advantage.

THE ECO-ADVANTAGE BOTTOM LINE

Here's our list of the top Eco-Tracking tools:

- AUDIO analysis
- Life Cycle Assessment
- Developing a core set of environmental indicators
- Establishing a materials database
- Comparative metrics to drive competition
- Environmental Management Systems
- Emergency procedures
- Partnerships

Chapter 8 Redesigning Your World

Green architect Bill McDonough likes to say that we humans should be humble. After all, it took us 5,000 years to put wheels on our luggage.

McDonough argues that we need to look at environmental issues in a new way. Eco-efficiency isn't good enough, he says. Too often, it's just making the wrong things more efficiently. McDonough sees vast opportunities to redesign products or even reconceive how and why we use them. He envisions a world where, after customers are done with them, products are broken down into either biological parts (like food waste or cotton) that can be safely disposed of, or technological parts that can reenter the industrial system and become new products. Only design that supports this vision, he says, will lead us toward sustainability in the deepest sense.

Redesigning products, processes, and even whole value chains is the second section of the Eco-Advantage Toolkit (see figure). To make real environmental gains and benefit from reduced waste and increased resource productivity, companies need to make fundamental changes to how they—and perhaps their customers and suppliers—do things.

Design is critical because so much of a product's environmental impact is firmly established in the design phase. As Timberland's Terry Kellogg put it, "Once you spec out a product, 90 percent of the footprint—in energy, water, chemicals, hazardous waste, you name it—is set." Design is where the rubber meets the road.

Eco-Advantage Toolkit—Redesign

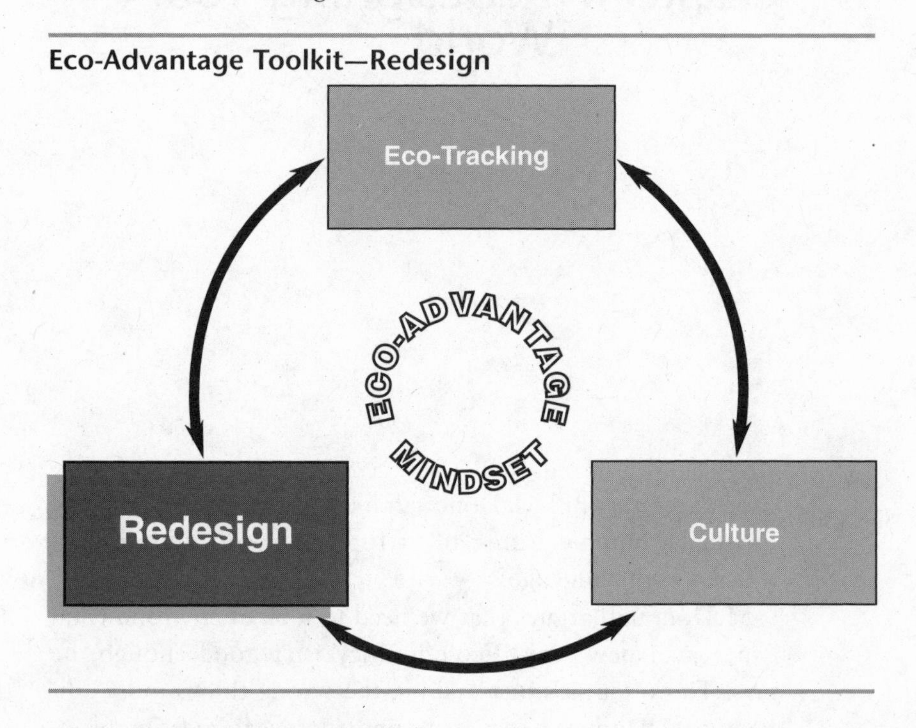

Pollution Prevention Hierarchy

McDonough and his partner, Michael Braungart, call their idealistic approach "cradle to cradle." For most companies, the state of the art in environmental thinking is more modest—and can be summed up with the slogan, "Reduce, Reuse, Recycle." As this simple guide suggests, the best pollution control option is to **reduce** the use of resources and eliminate waste. The next best option is to refurbish or **reuse** items. Then **recycle** what's left. As a very last resort, throw something out. Most companies are still working on integrating these three Rs into the production process. But why stop there?

Pollution Prevention Hierarchy

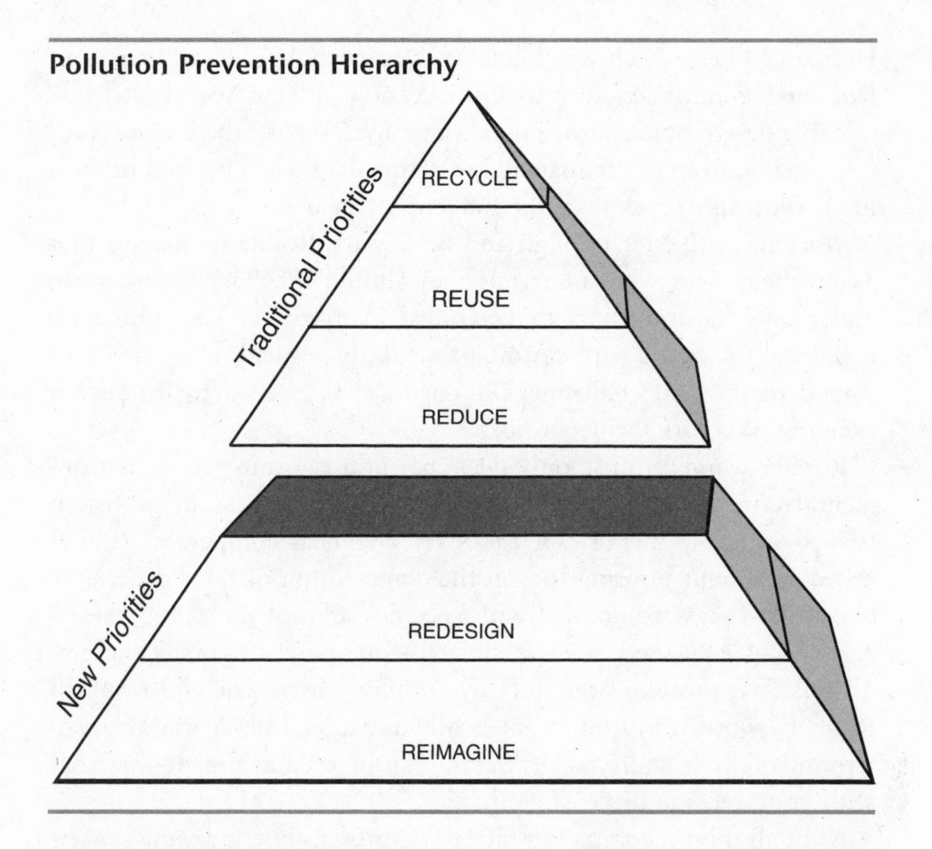

The pollution prevention hierarchy has two further levels (see figure). Before reducing, companies should explore ways to **redesign** what they do and how they do it. And even before that, they should try to **reimagine** their products or processes. Innovation is critical to 21st-century competitive advantage. The environmental lens can drive creative thinking and help companies find new opportunities to add value to their products and services and please customers. Just as companies have learned it's generally cheaper to reduce than to reuse, recycle, or throw out, now they are discovering that it is often more profitable to redesign and reimagine.

DESIGN FOR THE ENVIRONMENT

Swiss textile company Rohner Textil set out a decade ago to be a leader in the sustainability marketplace. Rohner saw a growing demand for environmentally safe fabrics from existing companies like

Herman Miller, which was launching eco-friendly product lines, and from new enterprises, such as Q Collection, a New York-based sustainable design home-furnishings company. To meet these market requirements, Rohner's managers knew they had to go beyond process improvements and explore the makeup of their products.

Working with McDonough and Braungart, Rohner managers first asked themselves what materials they should use. They chose wool and ramie, natural fibers that avoided many of the environmental problems associated with cotton, particularly pesticide use. Then they turned to the real challenge: the chemical dyes used to produce a rainbow of colors for decorators.

Rohner wanted to use only dyes that fit a rigorous set of environmental criteria: They could not cause cancer or contain persistent toxins or heavy metals. Of the sixty chemical companies Rohner asked to submit information on the composition of their dyes, only one, Ciba-Geigy, responded with specifics. Out of the 1,600 chemicals Ciba-Geigy used, a mere sixteen—one percent—made the cut. To this day, those are the only dyes Rohner uses. The end result of Rohner's work is a product they call Climatex, a biodegradable, environmentally friendly fabric. Scraps can be used as mulch—try that with your average piece of cloth.

As in all things, companies face trade-offs. Rohner's product meets all product standards except one: The dyes cannot be combined to make a pure black color. Sitting with CEO Albin Kaelin in Switzerland, we looked at swatches and could not tell the difference between "real" black and Rohner black. But of course, some interior designers can.

Design for the Environment, or DfE as insiders call it, is a catch-all name for initiatives like the one Rohner undertook. It's a systematic way to include environmental thinking in product and process designs from the get-go. DfE tries to minimize environmental downsides in input sourcing, production, and use.

Once a company has a feel for where environmental issues arise in the value chain, it can redesign to avoid those problems. For example, if end-of-life disposal is a problem, a product might be redesigned to make it easy and economical to recycle. Herman Miller's new Mirra Chair, the product of a rigorous DfE program, can be broken down in just fifteen minutes into a handful of parts that are

96 percent recyclable. In another design coup, the Mirra seat back uses a distinctive honeycomb design to give it the flexibility that chemical plasticizers would normally provide. This breakthrough allowed the Herman Miller team to design these chemicals out of the chair entirely.

Hitachi adopted a DfE strategy in its washing machine division to stay ahead of Japanese recycling laws. In redesigning its product to make disassembly easier, the company developed a process by which its washing machines could be made with just six screws. Not only did this new design facilitate recycling, the six-screw structure cut manufacturing time by 33 percent and significantly reduced the number of parts needed in inventory. Hitachi also discovered that the new washer required less service, so that customers got higher reliability and lower repair bills. Hitachi's efforts resulted in an environmentally preferable washing machine that's also a higher-value product with improved customer satisfaction, lower production costs, and reduced disposal costs.

CLOSING LOOPS AND INDUSTRIAL SYMBIOSIS

According to sustainability visionary Amory Lovins, "waste equals food." What he's talking about are the closed loops in nature. When a tree falls, it breaks down into dirt and provides nutrients for new plants. The waste becomes food for the next generation.

In the business setting, waste is costly. It often means you're squandering valuable inputs like metals, chemicals, or energy. WaveRiders have learned this lesson. They seek to recapture these resources by recycling waste water, reusing materials, and stripping valuable gases from exhaust. In doing all this, they reduce their footprint, improve resource productivity, and often save money.

Dow Chemical, for example, redesigned its process for scrubbing the hydrochloric acid used to make chlorinated organic compounds. The reengineered procedures, which recapture the acid, allowed the company to reduce its caustic waste by 6,000 tons per year. An investment of only $250,000 returned $2.4 million in annual savings on inputs and lower waste-disposal costs.

Herman Miller opened a waste-to-energy plant at one of its facilities. The company burns scrap fabric, generating over 10 percent of the facility's energy needs and all of its heating. Ironically, as Herman Miller has gotten more efficient, it has had to take scrap from competitors to keep the plant running at capacity. Not enough waste is an unusual problem, but not a bad one to have.

Process redesign can go far beyond recycling. Under pressure to reduce the use of fertilizers and pesticides that pollute nearby rivers and streams, the Dutch flower industry developed a new way to grow flowers in water and a material called rock wool. The closed-loop system recirculates the fertilizers and pesticides in the water, reducing the amount required and eliminating groundwater exposure. The new approach also lowers the risk of disease and narrows the variation in growing conditions, which improves the consistency and quality of the flowers. In addition, cutting and handling costs have fallen. As a result of its redesign initiative, Dutch flower growers have increased the value of their product, lowered the cost of inputs, reduced waste, and raised their resource productivity. All this translates into improved competitiveness and a world-leading market position for a country that lacks steady sunshine and faces stiff new competition.

The quintessential example of this sort of industrial ecology can be found in the town of Kalundborg, Denmark, where a series of companies are linked through a web of resource and waste flows some have dubbed "industrial symbiosis." At the center of the system is a power plant that throws off steam, heat, fly ash, and sludge from scrubbers. All of these waste products are useful inputs for other industries. A wallboard plant takes the sludge, a Novo Nordisk pharmaceutical plant and an oil refinery use the steam, and the town gets the excess heat. A Statoil refinery meanwhile sends the power plant waste water and gas. The web of connections also includes a local farm, a biotech plant, and a cement manufacturer. Everyone involved saves resources, time, and money.

REDESIGNING AND REIMAGINING SPACE: GREEN BUILDING

Eco-designed buildings are a big part of the Green Wave. Sustainable design principles can make buildings more energy efficient, lighter, and airier. Back in the 1970s, Yale's famous architectural historian

Vince Scully derided efforts to design solar heating into buildings as "plumbing," not architecture. No more. Today, designers take green building very seriously.

In its standards for the design and construction of environmentally sound buildings, the U.S. Green Building Council awards points for using recycled materials, incorporating energy efficiency into design, and other environmental benefits. Based on the total score, a building can be LEED (Leadership in Energy and Environmental Design) certified to Silver, Gold, or Platinum standards. Cities and states are getting into the act by passing laws mandating that all new government buildings must be built to LEED standards.

As they are in so many other instances, WaveRiders are very much on top of this trend. When we started our research, Herman Miller was the undisputed champ, with two of its buildings rated Gold—out of only eleven such buildings in the country. These buildings are bright, beautiful, and airy. But they've also proven to enhance worker productivity, which is where the Eco-Advantage comes in. Certified buildings are energy efficient, so they are cheaper to operate. One of Herman Miller's buildings has lowered utility costs per square foot by 41 percent compared with typical construction.

Now, green design is entering the mainstream. The rebuilt World Trade Center is being constructed to LEED standards. Bank of America's new headquarters in New York City aims to be perhaps the most eco-friendly office building in the world. Seeking LEED's top-of-the-line Platinum status, the new building will employ cutting-edge design to reduce energy and water use by half. Nokia is already one of the real leaders in this space. Its headquarters includes recycled materials and uses passive solar heating and lighting, providing an inspiring testament to the company's commitment to environmental values.

REDESIGNING AND GREENING THE SUPPLY CHAIN

Poor Kathie Lee Gifford. In 1996, she was just a TV personality who licensed her name for a line of clothing. Surely she did not expect to become a *cause célèbre* for NGOs campaigning against child labor. Yet to this day, the label sticks. Search the Internet for "Kathie Lee" and the top links will take you to sites about Regis Philbin or about sweatshops. Ouch.

Between Gifford's debacle and Nike's high-profile sweatshop problems, a perfect storm of supply chain questions came together. Where are your products from? Who makes them? How do they make them? All these questions and many more are now fair game, especially for big brands. No longer can companies say, "Well, that's not really *our* business." Nor can they claim they didn't know what was in their suppliers' products.

WaveRiders are building big supply chain audit programs to help redesign how their products are made. And they're trying to get to know *all* their suppliers. In today's networked, blog-happy world, a problem with any supplier, even a small one, can tarnish a big brand quickly.

But supplier audits are more than protection against public relations disasters. If a supplier lags on environmental or social performance, it usually signals more problems down the road. IBM, for example, has discovered that suppliers that do poorly on their audits are likely to fail later in other areas, like delivery time or quality.

IKEA's Supply Chain Redesign and the Staircase Model

Even IKEA, a true environmental pace-setter and socially responsible company, came under attack in the early 1990s for a series of supply-chain scandals. A Swedish documentary showed children working in Pakistan, reportedly on IKEA products. On the environmental front, the company took heat for using wood from endangered rain forests. These kinds of public relations nightmares can cost millions. When just one product, the Billy Bookcase, was found to have illegally high levels of formaldehyde, sales in Denmark alone dropped 25 percent.

The problems hit IKEA's management team hard. After much soul searching, they built one of the most impressive supplier audit systems in the world. Known as "The IKEA Way on Purchasing Home Furnishing Products," or more simply IWAY, this program is broad, deep, searching, and very well thought through.

The scale alone is incredible. About eighty employees working in the "trading offices" (purchasing department) around the world visit suppliers and rate them on social and environmental performance. Another eighteen employees are foresters by training and work exclusively on understanding where all the wood in IKEA's products comes from. This latter group is larger than the forestry departments of some *countries*. Combined, these auditors have done thousands of supplier checks.

IKEA shared its IWAY Evaluation Checklist with us, and it's an amazing array of questions covering compliance, emissions, waste, chemicals, safety, child labor, work conditions, forest sourcing, and ten other areas. "Checklist" is a big understatement for this fifteen-page form that urges auditors to get their hands dirty and can take days to complete. Just one example of the helpful "Notes to auditor" throughout, this one on hazardous waste: "Check that the procedure is working in reality. Check how empty containers and barrels are handled."

Another note encourages the auditor to "explain the IKEA philosophy. . . . check that the supplier understands the key environmental impacts and has started to measure and follow up." These lines demonstrate an important element of the process: IKEA does not just swoop in, give a rating, and leave. The company works closely with suppliers to bring them up to snuff.

IKEA suppliers from Mexico to Bangladesh have cleaned up their operations, spending millions on new equipment like waste-water treatment facilities. But IKEA also steps in and helps suppliers to reduce environmental impact directly. One Romanian furniture supplier, with loans from IKEA, invested in modern equipment including a new boiler, ventilation, and air filters, and installed a machine to turn briquettes from waste to energy and profit.

A core element of IWAY is what IKEA calls the "staircase model," which establishes four levels of achievement. Level 1 is basically unacceptable and means the supplier must have an action plan for reaching Level 2, the company's minimum standard. Every new supplier must go through an audit *before* delivering the first shipment. Level 3 is yet a higher standard, and Level 4 suppliers meet even stricter third-party standards such as the Forest Stewardship Council's certification, widely considered the toughest standard in sustainable forestry.

The staircase model is what elevates IKEA's supply chain audit above the vast majority of similar programs. If supplier checks exist at all, most ask basic questions about compliance with the law. It's a "CYA" business practice at best. IKEA's proactive stance, which demands continuous improvement, is a different animal. The company is pressuring suppliers to change how they operate, where they source wood from, how much they pay employees, and much more. Digging this deep represents real value-chain redesign.

So what does all this cost? It's a good question with a surprising answer: *No one knows*. "We've never put a budget together or calculated the costs," IKEA's Dan Brännström said. "It's rather fantastic." Don't misunderstand: IKEA cares deeply about costs. The company is famously value conscious—some might even say cheap. But the value to IKEA of lowering supply chain risk is so high that millions of dollars in employee salaries and time is seen as negligible in comparison. As IKEA's top CSR executive Thomas Bergmark told us, "For IWAY, there is *almost* no hurdle rate. . . . there's a risk by *not* doing it. It's not an option. . . . our brand name has real value."

To protect that value, IKEA even audits the auditors. As IKEA's Head of Compliance, Brännström, who came from running the intense store reviews on the retail side of the business, now makes sure the company meets the same high standards in the supply chain. On top of the internal audits, IKEA also periodically brings in third-party verifiers to review the whole process. It's a three-tiered system of checks that reduces supply chain risk about as much as anybody can.

Finally, if anyone in the organization has any doubts about how seriously IKEA takes this supply chain work, they can just ask the IWAY Council that oversees the whole operation. It's chaired by Anders Dahlvig, IKEA's CEO.

THE ECO-ADVANTAGE BOTTOM LINE

WaveRiders move beyond merely tracking environmental issues to change products, processes, their workplaces, and even their supply chains. To carry out their redesign efforts, they deploy various tools including:

- Design for the Environment (DfE)
- Closed-loop systems
- Industrial ecology
- Green building and LEED certification
- Supply chain audits

Chapter 9 Inspiring an Eco-Advantage Culture

In 1997, a Conoco oil tanker and a tug boat collided near Lake Charles, Louisiana, opening up a hundred-foot gash in the tanker. Few remember this accident today for one simple reason: Not a drop of oil was spilled. Conoco, then owned by DuPont, had invested in "double-hull" tankers years ahead of regulations. One hull was ripped open, but the second held. "That spill would have been larger than the *Exxon Valdez* without the double hulls," DuPont's VP for Environment Paul Tebo told us. "Conoco would have been gone."

What does this have to do with creating a corporate culture of environmental engagement? Quite a bit. In 1989, when green issues were still off the radar screen of most of corporate America, DuPont's visionary CEO Ed Woolard launched a Board-level Environmental Policy Committee and established an Environmental Leadership Council made up of senior executives who met every month. Woolard's green logic inspired fresh thinking across DuPont's diverse business portfolio.

Archie Dunham, Conoco's chairman, sat on the council and was, Tebo proudly reports, "a convert." With the zeal of the recently converted, Dunham and other DuPont executives in 1990 made an expensive commitment to build only double-hull ships. They believed that the reduction in risk to both the company and the environment was worth it, and were they ever right.

Many companies talk about having a culture of innovation, or they claim, "we put the customer first." But what does that mean, exactly? Often, not much. Too many slogans are attempts at quick fixes and panaceas for corporate ailments—empty words unless backed with actions and results.

3M has long claimed the innovation mantle and can prove it with countless new, leading-edge products that have kept the company growing for decades. The company's innovation record is no accident. 3M uses concrete organizational tools to stay fresh and drive new thinking. The company's famed "15 percent rule," for example, frees its engineers to spend up to 15 percent of their time on projects of their own choosing, no matter how wild.

Culture comes from more than just a high-minded mission statement or the words in a CEO's "all-hands" e-mail. It's built day in, day out with a conscious effort and incentives to shape people's behavior. Companies can encourage difficult-to-measure "soft" skills like creativity with "hard" rules and metrics. In *Good to Great*, Jim Collins talks about a proverbial flywheel that gets moving slowly but, with constant nudges in the right direction, gains speed and inertia. Building a corporate culture that promotes environmental thinking is a flywheel project.

The right tools help create the flywheel and get it spinning in the direction of Eco-Advantage. Everything from executing Green-to-Gold Plays to cultivating an Eco-Mindset to using Eco-Tracking and Redesign tools adds momentum. In this chapter, however, we focus on tools that create a culture that cultivates opportunities for strategic advantage.

The WaveRiders we've talked to use four basic culture-building tools:

- A vision, reinforced by stretch goals
- Practices that fold environmental thinking into every strategic decision

- Incentives for engagement with and accountability for the environmental agenda
- Communications aimed at both internal and external audiences.

Eco-Advantage Toolkit—Culture

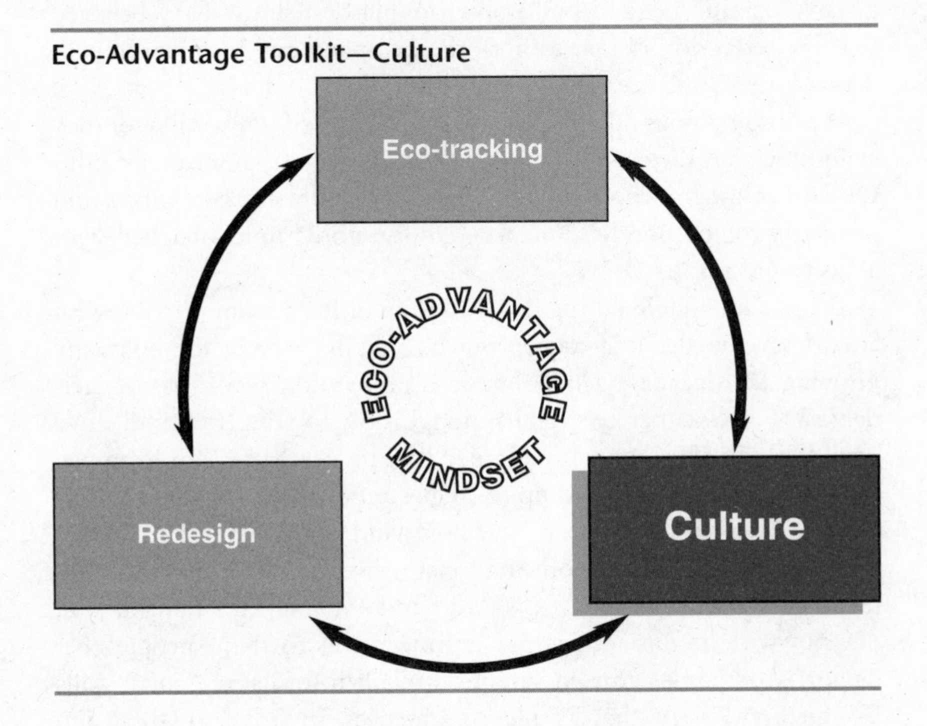

DREAM THE IMPOSSIBLE

When Katsuaki Watanabe took over as President of Toyota in 2005, he made it clear that developing environmentally friendly technologies would be his top priority—even ahead of safety, quality, and cost. If this weren't shocking enough, he also promised that his engineers would someday develop a car that could "cross the U.S. continent on one full tank of gas." Talk about a stretch goal!

Goals for an organization can be splashy like Toyota's or more concrete. Unilever, for example, set a fairly unsexy goal of zero liquid effluent from its seventy-six facilities in India. Watanabe himself set more realistic targets alongside his big 200-miles-per-gallon vision, including a sales goal of one million hybrid cars per year. Either way, fanciful or functional, stretch goals are a vital tool for provoking fresh thinking, promoting innovation, and building an Eco-Advantage.

As Harvard Business School's Michael Porter has argued, having to address tough environmental issues can be the spark a company needs to go beyond its comfort zone and find ways to innovate. Stretch goals drive creativity by asking the near impossible and demanding the reexamination of assumptions. They force everyone to search for new ways to meet old needs.

We're not talking about lofty statements like, "We seek to be an environmental leader in our industry." Visions help set a tone and establish priorities, but they don't provide concrete goals or much direction. Nor are we talking about incremental gains such as "use 5 percent less energy next year." We're focusing instead on big advances that present a challenge and might seem nearly impossible. The stretch goals we've seen animating corporate culture give life to the mindset principle "No is not an option."

Our research suggests that the sweet spot for stretch goals comes with targets that are clear and specific, yet far-reaching. Some are technically achievable but extremely challenging, like the 3M and Nike targets of reducing emissions of volatile organic compounds by 90 percent (both companies achieved it). Other stretch goals can seem purely symbolic, but in fact also drive execution. For years, DuPont has declared, "The goal is zero" when it comes to waste. While certainly an aspiration, zero packs a punch. It is specific and easy to visualize. If something is going out a pipe or into the garbage, you know you've missed the target. Furniture maker Herman Miller recently outlined its goals for the year 2020, which included zero waste and a 100 percent emission-free footprint. They've labeled these goals "Perfect Vision."

Stretch goals help to clarify a company's long-term direction and empower employees down the line to step up to the challenge. In 1999, Alcan set an ambitious greenhouse-gas reduction goal of 500,000 tons by 2004. Once unleashed, the company's engineers found 2.2 million tons to cut in only the first two years. Alcan's Dan Gagnier told us how management had missed the boat: "Executive management, no matter how well intentioned, was lagging behind operational management, who felt it was time the company did much better on these issues."

But here's an irony: While operational people often know more about what actually can be achieved, they should not be expected to set stretch goals. In the words of DuPont's Paul Tebo, "If you set goals by letting the organization tell you how to set them, you don't have any goals at all. . . . you just get what people know they can do."

Some WaveRiders are setting goals that help them cut environmental impacts beyond their own operations, up and down the value chain. Alcan, which consumes tremendous energy in making aluminum, has started talking about going "Beyond Zero." Here's the thinking: The production of every ton of aluminum creates about twelve tons of greenhouse gases (GHGs). If the transportation market uses more aluminum, lighter vehicles will burn less fuel and emit fewer GHGs. By Alcan's estimates, a ton of aluminum added to a car saves about twenty tons of GHGs over the life of the vehicle.

Critics would say that Alcan is just trying to get out of reducing its own emissions or, worse yet, that it's pushing this logic to sell more aluminum. Both may be true. There can be a fine line between broad thinking and issue avoidance. But thinking about how your products fit into the full value chain is the right approach—and often the best way to find opportunities your competitors may have missed, which is the definition of Eco-Advantage.

Going beyond zero can inspire everyone in the company to find new solutions and do what was once thought impossible. Over time, goals that might have been symbolic become achievable. Herman Miller now expects to reach its goals of zero landfill and 100 percent renewable energy. Our research consistently shows that tough standards often spark innovation.

WaveRiders ask their organizations what's achievable, then set stretch goals beyond that to unleash creativity, drive innovation, and build Eco-Advantage.

We heard the same thing almost word for word everywhere: "We didn't have a clue how we'd get there." Diving into the unknown can be scary. But this is why seemingly impossible goals like zero, or even beyond zero, are so important for breaking out of old ways of thinking. As Dawn Rittenhouse, DuPont's Director of Sustainable Development, told us, "We're not looking for continuous improvement. We're looking for strategic thinking and transformation."

GOING PUBLIC

Want to take goal-setting to a new level? A public declaration of environmental goals lights a fire under people's feet, but be careful. Setting public goals—and then missing them—can be painful. DuPont's Paul Tebo put it succinctly: *"Once you go public, it's no longer voluntary."*

ECO-ADVANTAGE DECISIONMAKING

In some companies, top executives decide nearly everything. Other corporations are highly decentralized and give people at all levels some latitude. But even with their different cultures, the WaveRiders we studied showed a few common trends in decisionmaking style that help them make tough environmental choices.

Environmental issues tend to be thorny and complicated. Traditional cost–benefit analysis often fails to include hard-to-see or hard-to-measure upsides in the calculus. But WaveRiders do *not* ignore intangible benefits. They know that payoffs may come down the road or in a form that traditional accounting has a hard time capturing, such as avoiding accidents, deflecting regulatory attention, and building goodwill with consumers and communities. Environmental costs and benefits are frequently diffuse or delayed, but they are no less real.

Rethinking the Hurdle Rate

Many possible corporate investments—new initiatives, product launches, marketing, and so on—compete for attention and money. To help decide among them, most companies require that any proposed investment yield a minimum return known as the "hurdle rate." By forcing everyone to justify proposed projects with hard numbers, hurdle rates provide a metric for evaluating investment alternatives and avoiding misallocation of scarce investment dollars. In the environmental realm, however, where intangible benefits may be undervalued and hard numbers difficult to establish, fixed hurdle rates can lead to wrong decisions.

3M, for instance, often slashes the hurdle rate for Pollution Prevention Pays (3P) projects from the corporate standard of 30 percent to only 10 percent. IKEA, too, gives special leeway to environmental investments. When deciding, say, whether or not to install solar panels on a store, the company allows a ten- to fifteen-year payback—much longer than for other investments.

In some cases, WaveRiders actually lower the hurdle rate to zero. One of 3M's manufacturing directors, Jim Omland, told us about two multimillion-dollar projects he approved without even calculating a hurdle rate. One factory cut the emissions of an air pollutant and another cleaned up the water going into ponds in Little Rock, Arkansas. Neither change was required by law, but making them kept the company ahead of stakeholder expectations. "Many issues fall into the 'you just do it when you can' category," Omland said. "If I don't deal with them now, I could face community or regulatory issues down the road."

Let's be clear. If a project provides an immediate payback in the usual commercial terms, then it's an eco-efficiency play and doesn't require a lower hurdle rate. The flexible hurdle rate is for projects *without* an immediate benefit in hard dollars. To green-light these projects, executives need a bit of slack and the courage of their convictions. What 3M and other companies are recognizing is that the intangible benefits of these projects—reducing risk, staying ahead of regulations, pleasing communities, enhancing employee morale—are often substantial, even if they are hard to measure.

Some companies use a formal process to identify lower hurdle rate projects. Unilever's capital investment process requires an environmental profile of an investment, which may trigger a lower rate. Others prefer a more informal process. McDonald's executive Bob Langert remarked, "For some investments, I don't have to show the normal hurdle rate. It's not formal, and you won't find it in writing anywhere, but I have the go-ahead for that sensibility."

Still other companies are beyond informal. They don't technically lower the hurdle rate but instead make strategic, directional changes without doing cost–benefit calculations. Take the flaring of gas from oil wells. BP and other oil companies realized that burning off the excess gas found in oil deposits was becoming unacceptable to many stakeholders. The practice contributed to global warming with no economic gain. So BP managers simply set a "no flaring" policy. BP's Chief Economist Peter Davies told us, "There's no point" in measuring the intangibles on such policy changes. Davies wasn't saying that intangibles are worth nothing. Rather, as an economist, he was acknowledging that they can't be measured precisely. What really matters in this case is that the strategic no-flaring decision was right for the long-term health of the brand and the company.

For all these business decisions, WaveRiders fold intangible benefits—such as lower risk, brand building, and reputation protection—into their investment calculations. The business logic of their choices is sound. Indeed, it's the traditional business tools that ignore key costs and benefits.

Pairing

During a recent redesign of its shoe boxes, Timberland eliminated 15 percent of the cardboard and reduced material costs by roughly the same amount. This was an obvious eco-efficiency win. But environmental executives also wanted to use 100 percent recycled cardboard in the new design. Post-consumer recycled content, however, costs more per pound. Luckily, the savings in material quantity offset the switch to a more expensive paper stock. But the savings could have been significantly higher had the company chosen to skip the expensive paper change.

Around the same time, Timberland was also exploring options for retrofitting two large distribution centers with low-energy lighting. The total cost was $600,000. Because of state subsidies, the retrofit of one facility in California would pay back in less than two years—a no-brainer. But retrofitting the other distribution center in Kentucky would take six or more years to hit payback, with an internal rate of return below the company's hurdle rate. Timberland could have chosen to do only the California retrofit.

In both cases, the managers "paired" the less desirable project with the no-brainer and presented them as a single proposal. The box redesign, paired with the switch to recycled content, would save only a little money—much less than the 15 percent of the box redesign alone. The two-site retrofit would have a three-year payback, exceeding the hurdle rate, but taking longer than the best project alone. Why make these choices? Because Timberland was keeping a big-picture focus on its corporate reputation and long-term value.

We saw similar "illogical" behavior at many companies. Herman Miller sorts twenty-six different waste materials and sends them to be recycled. For half of these materials, the company makes money, since the value of the recaptured scrap is higher than the cost of sorting and pick-up. Processing the other half loses money. But Herman Miller still sorts them all and breaks even on the total recycling program. And when they find ways to make more from the profitable materials, they bring additional money-losers into the mix.

WaveRiders such as Interface, McDonald's, and IKEA have used a similar form of pairing to increase their percentage of renewable energy. In particular, they plow the savings from energy efficiency projects into buying higher-priced green energy, keeping total energy costs flat while achieving broader environmental gains.

Pairing is a tool to keep good long-term decisions from being rejected for short-term considerations—an end run around the "green eyeshades" in accounting. By breaking even overall, managers reap the intangible upside benefits—bragging rights with stakeholders, enhanced brand and intangible value, and happier employees—at no net cost.

Leveraging the Invisible Hand

BP's Peter Davies told us that his environmental responsibilities were simple: Be certain the company makes good, rational decisions. For an economist, no decision-making system works better than one relying on markets. Economists love the magic of Adam Smith's "invisible hand." So we weren't surprised when Davies expressed his enthusiasm for BP's experimentation with an intracompany trading market for carbon.

The program started with targets on greenhouse gases for each of twelve business units. The overarching goal was to reduce company emissions by one percent per year or more. To meet this goal, the company established an internal trading system that lets business units buy or sell emissions reductions. If one business could cut greenhouse gases more than its target, it could sell the additional tons of carbon to other divisions. BP's business units traded two million tons at a shadow price of about $5 per ton.

Shell set up a similar internal carbon trading market, gearing the company up to meet its Kyoto Protocol obligations in Europe. Both BP and Shell recognize the limits of the intracompany carbon game. They don't worry about whether they've got the carbon price exactly right. The management teams understand that internal trading is simply a tool to draw attention to emissions and highlight opportunities to reduce them. Garth Edward, Shell's Trading Manager for Environmental Products, buys and sells commodities in actual global markets. The internal trading system, he says, is not real, but "it's useful as a flag-raising exercise."

BP also recognizes that the goal is to focus business leaders on the issue. As BP's Chris Mottershead told us, "In business, you can't focus on 10,000 things, but can only do about three things well. With this tool, reducing the impact of our operations on climate change became one of those three."

The idea of casting a spotlight on pollution costs and imposing an internal "tax" on environmental harms has many variations. In 1995, after CEO Albin Kaelin declared, "The traditional accounting system is not telling the truth," Rohner Textil put a tax on itself for its carbon emissions. Herman Miller similarly buys "green tags"—renewable energy credits that support clean energy development—as a way to get managers focused on the impacts of higher energy costs that they likely will face in a carbon-constrained world.

One WaveRider even funded some of its most forward-looking environmental work out of a special kitty controlled by the Board of Directors. The money supported a pilot program on Design for the Environment and other nontraditional projects. In effect, the Board subsidized the work and artificially lowered the price of these investments, thereby spurring environmental gains, creative thinking, and some risk-taking that might otherwise not have happened.

A few companies are even using environmental charges to help consumers and employees make better decisions in their own lives. IKEA put a small tax on plastic bags—and donated the proceeds to charity—to help customers see that the *environmental* cost for using the bag was not zero. Hyperion Software and Timberland have both started programs to help employees buy hybrid cars—to the tune of a $5,000 rebate in Hyperion's case.

In the end, internal trading or taxing mechanisms change price signals. Emissions of greenhouse gases cost the world something, but they're currently priced at zero for those who release them. Market mechanisms help companies "internalize the externalities" and move the price closer to the cost to society of the environmental harm. And as regulatory mechanisms kick in, the companies that have moved ahead on their own will be better positioned for success under the new competitive conditions.

BROAD ENGAGEMENT

The CEO launches a new initiative. Everyone scrambles. The CEO leaves or gets sidetracked by other priorities. The initiative peters out and finally dies. Does this "flavor of the month" management style sound familiar?

It's no different with Eco-Advantage initiatives. While CEO leadership is required to make the environment a top priority, these efforts will not get very far if *only* the CEO is engaged. Executives of all stripes—most importantly, middle management—need to own the issue.

All business unit heads at Shell, for example, write and sign an "assurance letter" vowing their compliance with the company's sustainability priorities (somewhat like the CEO and CFO signatures on financial statements required by the Sarbanes-Oxley law).

WaveRiders drive real environmental engagement by:

- Placing ownership for environmental strategy with management at an operational level.
- Cross-fertilizing executives between line operations and environmental positions.
- Establishing incentives for environmental success based on clear metrics.

ECO-ADVANTAGE CHAMPIONS

At a number of companies, valiant eco-champions work the system to get environmental initiatives moving. Some even have CEO support. But those in the environmental department cannot be the sole source of environmental action. For fresh thinking to take hold and generate Eco-Advantage, the practice of looking at choices through an environmental lens has to be embedded throughout the organization. Relying *only* on a champion is a doomed strategy. Real success requires engagement from the top of the organization to the bottom.

Ownership

A sprawling committee that includes 400 of 6,000 employees would not seem like the most efficient organizational tool. But furniture maker Herman Miller's Environmental Quality Action Team (EQAT) works extremely well. It's the best tool we saw for engaging all levels of employees in the goal of environmental stewardship and the challenge of building Eco-Advantage.

Begun in 1989 with a few people working on lowering the environmental impact of manufacturing, the EQAT has grown organically over time. It's now a loose matrix of nine subcommittees covering topics as diverse as design, transportation, and green marketing. The EQAT has minimal hierarchy, but an environmental executive informally coordinates the whole structure, and a handful of managers form a core team. This group guides without commanding, while the subteams set and measure progress against stretch goals. As an engagement tool, it's powerful. When a team needs buy-in, it invites people with key roles throughout the organization to join the fray.

As the company's Design for Environment (DfE) initiative evolved, the EQAT asked the Senior Vice President for Purchasing, Drew Schramm, to join the team. After all, how could they design products with environmental considerations without involving the person in charge of buying everything that goes into those products?

Soon Schramm faced a tough choice with serious environmental implications. His financial goals for the year were tight: Cut $25 million in costs. His team found a less expensive replacement for one commonly used part. The new option would save the company $1 million, but it contained polyvinyl chloride (PVC). And unfortunately, the EQAT's DfE team had set a goal of using no PVC. In the long run, as potential regulations tightened or customer preferences shifted away from harmful materials, it would be better for the company to avoid PVC. So Schramm said "No thank you" to the $1 million savings, took his decision to the president, and together they agreed it was the best course. Schramm had to find the savings elsewhere, which he did.

Now imagine that same scenario if Schramm had not been part of the team that had actually chosen the no-PVC goal. The conversation might have gone something like this:

> EQAT members: "Hey, Drew, how about you give up that $1 million in savings so we can make our PVC goal?"
> Drew: "Yeah, sure, I'll get right on that."

Maybe that's a bit unfair—Schramm might have made the same choice. But clearly it was easier for him to make the call and to make the case to his boss as an EQAT team member. He owned the environmental goal as much as anyone.

The EQAT is a strange animal in almost every regard, size and reach being only the most obvious. It has a very loose reporting relationship, and it's a great example of empowerment. The team escalates an issue when they think it's necessary, and million-dollar decisions meet the informal threshold. But they have no fixed reporting schedule. Herman Miller's President and CEO, Brian Walker, told us he'd like to hear from the EQAT more. (How often does an executive want more reports from a committee?) But, he said, "Why tinker with something that works?"

High-level committees focused on environmental or sustainability issues are becoming more common. Starbucks created an Environmental Footprint Team with executives from critical areas such as transportation, purchasing, and store operations, but also reaching out to human resources, public affairs, and legal. The team meets quarterly to discuss progress against their goals. Very senior-level teams are cropping up as well. European chemical company BASF established a Sustainability Council made up of Board members and seven division presidents.

Committees are no cure-all, as every executive knows. Often they mean more meetings for busy people. They can also easily backfire by making members tune out. But committees can be a powerful start. Better still, as the culture of the company embraces environmental thinking, senior executive committees tend to become less important. DuPont has moved away from the monthly Environmental Leadership Committee meetings, and now has one Sustainable Growth Council led by the chairman. They meet only three times per year, which could be a sign of lessening commitment. But for DuPont, it indicates that they've moved environmental thinking deep enough into the company to reduce the need for top-down management.

Companies can also ratchet up their efforts by giving up-and-coming executives real responsibility for company-wide, cross-cutting programs. In 2004, Hector Ruiz, CEO of chipmaker AMD, wanted to engage his management team in environmental and social issues. So he asked a mix of senior and junior managers each to shepherd a sustainability issue through the organization.

This Executive Stewardship program leverages the EHS staff and builds "bench strength" by testing talented employees on important projects. Most importantly, it moves the ball forward on important initiatives that might otherwise falter for lack of support. Executive stewards lend senior support, help eliminate organizational barriers, and are directly accountable to Ruiz. As senior company officials, the stewards bring clout to the environmental agenda. And by giving them responsibility for action, Ruiz can be assured that the stewards dig in hard.

Spreading the management focus on environmental and social issues makes sense. Timberland took aspects of Corporate Social Responsibility and divided them among three senior executives who report to the Chief Operating Officer. Timberland's Terry Kellogg told us, "Our CEO is still the keeper of the CSR flame, but now we have three of seven top executives carrying the voice."

TOUGH LOVE: GE'S SESSION E

At GE, plant managers and operations leaders in every line of business, in every region of the world, must explain their plant's environmental, health, and safety performance to a panel of EHS executives, the head of his or her business unit, and a room full of peers. In these annual meetings, called "Session E," the plant manager makes the presentation, *not* the plant's environmental manager.

At Session E—an off-shoot of the GE's more famous Session C, which helps the company identify top talent—managers are praised for outstanding performance and exchange best practices with their peers. But they also get direct feedback about their failures. "In thirty years in business, I've never been so humiliated," said one veteran factory manager after a dreadful performance at Session E.

Another manager had failed to achieve closure on 78 percent of environmental and safety issues left over from the previous year's review. First came a public dressing down in Session E. A few days later, the Session E Letter that went out to all participants summarized the meeting and *again* addressed this manager's shortcomings. This is peer pressure at its worst—or best. Nothing says "you're accountable" like the specter of falling flat on your face in public.

GE's tough love seems harsh, but it works. And GE managers aren't left to sink or swim on their own. The company's EHS team provides a range of tools to help managers improve their performance. In fact, the "humiliated" factory manager turned his plant into a world-class EHS facility over the next few years.

Cross-Fertilization

Before taking over as IKEA's top sustainability executive, Thomas Bergmark ran the dining and home office furniture business. 3M's top environmental official, Kathy Reed, spent twenty years in line operations. DuPont's Paul Tebo was running the company's petrochemicals business when he was tapped for the environment job. Notice a trend?

WaveRiders make a point of bringing line expertise into critical environmental roles. Career environmental professionals are talented and vital to the organization. But when environmental executives make demands, line management is inclined to dismiss them with a quick, "You don't understand the financial pressures we're under." That retort doesn't fly with managers like Bergmark, Reed, or Tebo—they've been there.

One caveat: Companies that park a senior executive in the environmental role as a last stop before retirement have problems. The environmental management agenda is far too complicated and demanding to be handled by someone counting the days until he can collect his pension.

Experienced executives from operations have the connections and credibility to push the environmental agenda. They know the business imperatives and are often better positioned to spot opportunities for generating Eco-Advantage.

Cross-fertilization is a powerful tool at all levels. When Bergmark needed an environmental manager to act as the go-between with store operations, he brought in Nicole Schneider, a long-time operational manager from one of Germany's biggest IKEA stores. Why pick her and not an environmental person? "The key is to link environmental affairs to the business and get out of the 'greeny' corner," Schneider told us. "I know how I thought when I was there."

Cross-fertilization also means getting people who normally would not speak to one another in the same room. When Toyota wanted to break the mold and develop its next generation car—the one that became the incredibly successful Prius—top managers led brainstorming meetings with people from across the company. Herman Miller, while trying to meet its tough zero-landfill goal, put janitors together with engineers to discuss where the company generated waste.

For some companies with the right culture, the ultimate in cross-fertilization is to make environmental thinking a deep part of *many* people's jobs. Then you don't need special meetings to merge different perspectives. Perhaps Nike is the furthest along this path. Of about 25,000 employees, thousands have some responsibility for sustainability issues, and many dozens of these have something in their title to that effect. For example, the company designers follow the principle of "Power of One," meaning they must find one design element per product season that reflects sustainability priorities. Nike credits this diffusion of thinking to the fact that the company started as a design and marketing firm, without the in-house manufacturing operations that require traditional compliance-oriented executives. When the company took on sustainability and responsibility for its value chain, it found ways to engage people from all over the organization. The environmental ethos happened organically.

Cross-fertilization and ownership are about integration. Companies find opportunities to get ahead of the competition when *everyone* looks at the business through an environmental lens. WaveRiders ask all employees to help find Eco-Advantage.

Incentives: Compensation, Careers, and Kudos

If what gets measured gets managed, you can bet that what people get *paid for* (or promoted for) will get managed even better. Let's face it: When most managers hear the word "green," they think dollars, not environment. So it would seem logical that WaveRiders would put environmental performance into executive bonuses and compensation. In fact, some do, but not all.

Sure, WaveRiders believe that the environment must become a nonnegotiable part of everyone's job, just as safety and quality did in the 1980s and 1990s. But each company acts on this belief in its own way. In many companies, environmental results are now part of the key performance indicators around which executive evaluations revolve. In others, environmental performance factors into bonus calculations. For still others, environmental performance becomes ingrained in the company culture, affecting who gets promoted and even who survives in the company.

In 1999, Northeast Utilities found itself reeling from a series of guilty pleas over its failure to uphold pollution laws. So the company refocused attention by making environmental results 20 percent of every executive's performance evaluation and bonus calculation. That single action sent a wake-up call that was sorely needed.

The reasons for turning bonuses a shade greener are not always so stark. Many companies just believe that the environment is a legitimate priority to encourage. At Shell, performance against key CSR indicators—such as greenhouse gas emissions, oil spills, injuries, and diversity—determines up to 25 percent of executive bonuses. SC Johnson includes in its managers' bonus calculation their product-line Greenlist scores, which rate toxicity and environmental impact. And Chiquita ties bonuses for farm managers to their compliance with the Rainforest Alliance's banana certification program.

Some of the companies we spoke with had little *visible* relationship between pay and environmental performance. If environmental thinking is embedded deep in the corporate mission, monetary incentives are redundant. It becomes unacceptable to operate any other way, and your job is at stake if you don't get it. At 3M, top managers know that environmental care is a baseline obligation. "You don't get an executive position if you don't get the values," Kathy Reed told us. "So if you want a career, you have to take environmental issues into account . . . *no incentive will save your job*."

Even if no money is involved, making the environment a key part of performance reviews sends a clear message. Dick Hunter, Dell Computer's corporate VP for manufacturing and distribution, has two-hour operational reviews with Chairman Michael Dell and CEO Kevin Rollins. The conversation now *starts* with safety and environment. In a cultural trickle-down, Hunter starts all his own staff reviews that way, as well.

The repercussions at some companies are very real. After IKEA built its IWAY supplier audit program, some managers didn't seem to buy in and manage supplier performance closely. So IKEA, euphemistically, "changed a few area trading managers." A company that removes people for missing the environmental boat doesn't need to do a whole lot more.

KUDOS

Finally, we'll say just a few words about a very obvious culture builder and behavior influencer: awards programs. Most companies have some sort of recognition—from plaques to dinners to hard cash—for employees who come up with ideas to save money or develop new products.

WaveRiders develop programs aimed squarely at environmental issues. 3M has a range of awards, most connected to its Pollution Prevention Pays program. It's not quite the Oscars, but "Most Hazardous Waste Prevented" is a nice honor to get at a special lunch with your coworkers. A few hundred 3P awards are presented each year. Out of this group of winners, a special few receive the Chairman's Environmental Leadership Award, including being flown to Minneapolis for a big dinner event with the top brass.

FedEx Kinko's launched an Environmental Branch of the Year program that honors the top few stores on environmental performance. And Timberland started a significant cash award program for large-scale initiatives, such as a major switch to renewable energy at the company's distribution center in the Netherlands. The Carden Welsh Award is named for a beloved former executive, which gives it a personal touch. The winner receives $1,000 in cash and $1,000 to donate to a favorite charity.

STARBUCKS MISSION REVIEW

How do you make sure your operations are living up to your ideals? Empower employees to check up on you. At Starbucks, *any* employee can ask for an explanation if something seems to violate one of the company's six guiding principles. In one case, a few employees noticed that shipments from a roasting plant were coming in with boxes that were not full. Sensing the amount of paper and fuel wasted, they questioned whether half-filled boxes really fit the company's principle on environmental responsibility. The plant upgraded its computer system to ensure fuller shipments. Giving employees the ability to question anything is a powerful engagement tool.

TELL COMPELLING STORIES

Years ago, an English teacher told one of us (Andrew) something that stuck: The world has only eight archetypal stories. And the Old Testament pretty much hits them all: boy meets girl (Adam and Eve), rivalry (Cain and Abel), rebellion against authority (apple anyone?), and so on.

Companies also tell stories, most of them built along familiar themes. Drink our beer, for example, and sexy women will love you. But when it comes to telling a company's environmental story, things get more complicated. Stakeholders listen carefully to green pitches, and if the facts are wrong or the assertions off base, they'll be sure to let you know, sometimes in a very loud way. Being transparent and telling the truth aren't exactly the hallmarks of traditional marketing, but in eco-marketing, you had better be clear . . . and right.

We covered some of this communications terrain when we explored the Green-to-Gold Plays that deal with marketing and intangibles. Here, we focus on particular "nuts-and-bolts" issues. We'll break up the listening audience into two core groups—external (especially communities, regulators, and NGOs) and internal (essentially employees).

External Audiences

Companies can use communications as both a shield and a sword. Touting the green qualities of a business can ensure that NGOs, regulators, the media, and ultimately, the public see the company as a responsible corporate citizen. Telling customers about the green attributes of a particular product also can set you apart in the marketplace so long as the benefits you claim are really there.

3M developed a simple process for managing these issues. Its Environmental Marketing Claims Committee is a cross-functional group that draws from public affairs, EHS and product responsibility, and legal affairs. It checks every marketing claim, from ads to packaging, and thinks about how a particular claim will fly in the marketplace. The point, the team says, is partly legal but mainly about reputation: "We want to be known as a company that's as good as its word."

More general communications about the company's environmental qualities can take a number of forms. Many companies now issue an annual environmental report or a sustainability or CSR report that addresses both environmental and social issues. Some just post relevant information on their websites. Which approach works best? It really does not matter—what's important is the content.

A good environmental report should discuss the important aspects of the company's footprint. It should use quantitative metrics and cover core issues such as air emissions, water pollution, hazardous waste disposal, energy consumption, greenhouse gas emissions, and notices of legal violations.

Failure to review all aspects of the company's footprint, most of which involve public information anyway, signals to stakeholders that a company has something to hide. Deal honestly with bad news—chemical spills, legal penalties, or pollution trends that are going up rather than down. Those who share shortcomings get to tell the story their own way and explain in their own words how the problem is being addressed. After its legal troubles in the 1990s, Northeast Utilities issued an environmental report with a sober letter from then-CEO Mike Morris on its cover with a blunt message acknowledging serious failures. Morris (now CEO of American Electric Power) committed Northeast Utilities to improved environmental performance. In the intervening decade, the company has come a long way.

Remember, too, that all businesses today operate in an ultratransparent world. When NGOs or the media find you trying to sweep an environmental breakdown under the rug—and they will—you can be sure that they won't tell the story sympathetically. Multinational companies are especially vulnerable, given their size and exposure. Even smaller companies can't keep bad news hidden in the face of local NGOs and an ever-expanding corps of watchdogs. Bloggers and other self-appointed supervisors of the public interest pretty much assure that nothing remains a secret for long.

Environmental reporting can and should also include good news—employee beach or river clean-ups, partnerships with conservation organizations, or profiles of employees who volunteer with community groups. Just make sure you don't rely on these feel-good stories to blunt the demand for cold, hard facts about the company's environmental performance. The days where a company could talk only about the good news and release a report full of glossy pictures of happy animals are long gone.

No definitive model will tell you what issues a good report should address or how best to present environmental information. The Global Reporting Initiative has tried to develop a template for sustainability reporting, but its early efforts have generated cumbersome forms of limited use in communicating with the general public. In the absence of a template, remember that good charts and graphs are key to any report today. Most companies find it useful to show trends—several years of past performance—and to track results against past and future goals and targets. And in case it's not obvious, all reports and recent data should be made available on the company's website. (Our website, www.eco-advantage.com, links to some award-winning reports and best practices.)

A final point: These reports, especially the web versions, offer an opportunity to *listen* as well as talk. Storytelling should be a two-way street. Knowing what others think of your performance and policy choices is very valuable. Shell's Tell Shell program, for example, has elicited wide-ranging feedback for the company, helping it to reframe its image and priorities. Shell now quotes its critics and fans alike in its reports. Likewise, SC Johnson invites experts and customers around the world to engage with them through a sustainability-focused e-mail exchange.

SUCCESSFUL ENVIRONMENTAL REPORTING

Tell the truth. Be upfront about bad news. Transparency is a powerful force behind the Green Wave, and expectations rise daily. Be clear about what metrics the company does and doesn't track, and show them in relative and absolute terms. Reporting on greenhouse gas emissions per unit of revenue is fine, but adjusting the denominators doesn't fool serious readers. And you're likely to annoy them if the overall total is growing rapidly. Environmental reports are a powerful tool to build trust with all stakeholders. Don't waste the opportunity.

Talking to the investor community about environmental initiatives takes particular skill. Annual reports and quarterly analyst calls, the normal channels for communicating with Wall Street, tend to focus on one story: quarterly earnings growth. Long-term sustainability strategies are a tough sell in a forum so heavily built around short-term gains. Interface's Chairman Ray Anderson told us, "It was seven years before I talked about anything to shareholders besides our cost and waste reduction initiatives." DuPont's CFO, Gary Pfeiffer, makes the point that companies wanting to talk about sustainability must talk Wall Street's language as well: "Wall Street will never accept 'in 100 years you'll love us.' . . . they want to know how we'll get there."

To keep the analysts happy and listening, sustainability initiatives need to be broken into bite-sized chunks and explained in strategic terms. GE's Jeffrey Immelt doesn't sell ecomagination as a feel-good campaign to make GE seem eco-friendly. He talks about fast-growing markets for environmental goods and services like fuel-efficient jet engines and better water treatment technologies. Wall Streeters find it infinitely easier to get their heads around a sales growth message.

The Internal Audience: Telling Employees Stories

Employees' identities are powerfully shaped by a company's reputation. Almost everyone wants to work for a company that they can feel good about. In our knowledge-based economy, the most prized workers are highly mobile and often highly sensitive to their employer's environmental performance.

Time and again, we heard from employees who responded to their companies' own CSR reports by saying, "I didn't know we did all this, and I'm so proud." The stories of emissions reduced, energy saved, or communities supported are heart-warming and build real loyalty. Even the failures are important. Employees know many of those stories anyway, but seeing their management own up to shortcomings builds trust.

Feel-good aspects aside, environmental reporting can have significant tangible benefits. First, a public report is a way to put a stake in the ground and signal to all employees what issues to focus on. The Shell Report acts as a form of leverage on employees and helps the company "manage the underlying performance," according to Shell's Mark Weintraub. Since the report is sent to 2 million people, including all 120,000 employees, executives take notice if their businesses are mentioned. The commitment to sustainable development is measured publicly in that report. No employee wants to look bad.

Second, a good report helps employees understand why environmental concerns or sustainable development are a corporate priority and why it's good for business. Shell, Weintraub says, tries to "push beyond the nice happy stories where we save whales to the business case." Creating opportunities to find Eco-Advantage requires everyone in the company to understand the value of making environmental thinking part of everyday strategy.

Finally, the report is a fantastic way to share learning across an organization. Gathering the data often reveals which facilities are leaders and which are laggards—and *why*. Companies also discover strategies and environmental wins they didn't know about and can now share more broadly. Weintraub observes that Shell uses its report as a mirror on the organization to see what's working and what isn't.

BEYOND REPORTING

Sharing best practices can be a dry exercise, or it can inspire others to follow. A compelling story opens minds. One of GrupoNueva's subsidiaries, Amanco, found a sizable $6 million in savings after a concerted eco-efficiency initiative uncovered fresh ways to reduce waste and cut resource use. Sharing these results with other divisions proved to be a powerful door-opener for Maria Emilia Correa, the VP for Environmental and Social Responsibility. "Successful stories get people interested," she said. "The $6 million savings for Amanco is a magic wand—it immediately opens eyes."

Companies need a range of tools to convey the environmental message. Some are experimenting with multimedia presentations, highly visible environmental bonuses, end-of-year awards that highlight employee efforts, intranets, and much more. Herman Miller and FedEx Kinko's have used company-wide videos as a way to tell stories about what each company is doing on the environmental front. DuPont created an Active Energy Leadership Team that swoops in on businesses to share best practices in energy management. BP shares information over its intranet, including case studies and stories on reducing carbon dioxide emissions on an oil platform or ways to reduce gas flaring. No single answer is right. What's best is whatever reaches employees and shows them how they can improve their operations.

ECO-TRAINING

Building an environmental mindset, identifying opportunities for Green-to-Gold Plays, and developing and using tools that create Eco-Advantage—none of these things comes easily. Like learning any other skill set, mastering these requires educating people up and down the line. We highlight here three kinds of training programs WaveRiders use to help create an Eco-Advantage culture:

- Training on focused topics like regulatory compliance or eco-efficiency

- "Take it home with you" informal education to raise general knowledge of environmental issues
- Executive-level, big-picture programs on sustainability.

In the first category, WaveRiders have developed significant training modules on specific environmental topics. Alcan has come up with a program it calls EHS First, a central vision for the company with the training to back up its importance. As Simon Laddychuk, who led the team that developed this initiative, told us, a key goal was to drive the philosophy into the core business. "All employees, and all of the top 800 managers, even Travis [the CEO], have gone through a four-day training module on environmental, health, and safety issues." When Alcan bought another large manufacturing company (Pechiney), it budgeted $20 million to harmonize the two companies' approaches to EHS and make sure everyone was at the Alcan EHS standard.

The second category, building employee interest in and knowledge about environmental issues, can be informal and fun. At Rohner Textil, employees loved the company's sustainability efforts so much they demanded more. One group wanted to know why the company wasn't using solar panels on the roof. CEO Albin Kaelin knew that solar energy wasn't a viable power option for Rohner, especially in Switzerland with its limited hours of sun each year. But why be the bearer of bad news? Instead, at the annual summer barbeque, Kaelin planned a demonstration. The refrigerator would be powered by solar panels. Everything was fine until night fell and the beer and wine got warm. Point taken. In another instance, Kaelin held a two-day seminar with no energy and little water. He had everyone cook with wood to make the point that resources are scarce. In both cases, Kaelin was handing out the same advice editors give writers: Show, don't tell.

Maybe Rohner does take its environmental consciousness-building a bit far. But many companies use newsletters and other informal methods to share information and generally raise everyone's level of knowledge. Clif Bar produces a newsletter called "Moving Toward Sustainability," which teaches employees about important issues for the business, like the benefits of organic farming. The company also uses a more informal e-mail series called "Notes from Your Company Ecologist" to cover issues that touch people's everyday lives, like the environmental impacts of dry-cleaning chemicals.

At the executive level, education and continuous knowledge refinement is even more important. A growing number of companies are making sure that their next generation of leaders is up to speed on sustainability and why it matters to the company. Unilever Co-CEOs Niall Fitzgerald and Antony Burgmans took the company's top 200 managers on a retreat to the Costa Rican jungle. The essence of the get-away was team building, but the environmental setting helped set a tone for the hard thinking required about the company's future.

Alcan's executive training, Laddychuk told us, is a "combination of technical and softer issues. . . . it's a translation of values into everyday practice—it's about a mindset." Similarly, Northeast Utilities developed an innovative training program for middle managers centered on "hard choices." Rather than dodging the reality that line managers have competing pressures that make environmental goals seem secondary, the company tackles these issues head-on and talks about how to strike a balance.

THE ECO-ADVANTAGE BOTTOM LINE

To build an Eco-Advantage culture, leading companies use many approaches:

- Stretch goals
- Decision-making that expressly accommodates environmental issues
- "Pairing" environment-related projects or adjusting hurdle rates to reflect undercounted intangible value gains
- Internal markets to highlight hidden environmental costs
- Stewardship committees to promote executive engagement
- Placing environmental ownership on operational officials
- Environmental performance reviews and clear accountability for results
- Cross-fertilization between environmental officials and those with line responsibilities
- Eco-based bonuses and awards
- Including environmental elements among key performance indicators
- Environmental or sustainability reports
- Real-time eco-information websites
- Employee authorization to question actions inconsistent with environmental commitment
- Eco-training

Part Four **Putting It All Together**

We've now covered all the pieces of a successful environmental strategy. Once a company has developed the basic tools for focusing the environmental lens on strategy and begun to instill the Eco-Advantage Mindset, everything will run smoothly, right? Well, not quite. As we've said before, environmental initiatives quite often go awry.

In Chapter 10, we review a core set of problems to avoid. Our research suggests that companies can get tripped up by their own organizations or from misunderstanding the nature of the challenge, among other things. But we have found some solutions, and we also review those.

As a stack of business bestsellers will proclaim, execution is everything. Likewise, gaining an Eco-Advantage is not just a thought exercise. Turning thinking to action, however, requires real work. In Chapter 11, we provide a plan of attack with specific guidance on short-, medium-, and long-term actions.

Finally, in Chapter 12 we provide an overview of the

Green-to-Gold take on environmental strategy. For readers who've read the whole book carefully, this chapter will be a review and might merit only a quick skimming. For those in a hurry and who've been skimming along up to now, here's a good place to slow down. In Chapter 12, we bring all the key concepts together in a concise way.

Chapter 10 Why Environmental Initiatives Fail

Ford's factory on the Rouge River in Dearborn, Michigan, has a storied history. Built before the Great Depression, it held 100,000 workers in its heyday, but in recent decades, the plant was in decline. To signal a rebirth of the company, Ford Chairman Bill Ford, Jr., decided to remake his great-grandfather's grandest factory. The redesign would also show the company's environmental commitment and "transform a 20th-century industrial icon into a model of 21st-century sustainable manufacturing."

Ford hired star green designer William McDonough to rethink the giant plant entirely. After a $2-billion overhaul, the new factory promised to be a paragon of efficiency and environmentally sound design, including a ten-acre "living roof" of grass that captures rainwater and reduces the energy requirements of the building. The site includes solar panels, fuel cells, and constructed wetlands. Sounds great, right?

Well, yes and no. The green factory may indeed be a marvel. It doesn't, however, begin to address the real environmental issue at Ford: gas-guzzling vehicles that contribute to climate change and local air pollution. From a life-cycle perspective, Ford's environmental footprint falls most heavily in the product use phase. Environmentalists won't declare Ford a green company until it makes real and sustained improvements in fuel efficiency and greenhouse gas emissions from its vehicles.

So was the factory redesign a failure? Not entirely. But this initiative accomplished far less than Bill Ford hoped it would. The lesson is that no company can afford to focus narrowly on its *own* issues—in this case, manufacturing processes—if this means ignoring the big concerns in its value chain. Failing to take seriously the reality of the extended producer responsibility movement stands as number one on the list of fundamental reasons environmental initiatives fail, as revealed by our research.

A quick review of the literature on green business over the last ten to fifteen years might lead you to think that life was all wine and roses. The published books, articles, and case studies almost exclusively tell stories of environmentally driven initiatives that paid off. A casual reader might even be tempted to believe that corporate environmental strategy offers win-win outcomes all the time.

The reality is much more checkered. In many cases, eco-efficiency efforts and other environmental investments do pay off. But lots of initiatives fall flat. Some don't deliver the promised environmental gains. Others don't work from an economic perspective. Some fail on both accounts.

There's nothing new about that, of course. New product launches come up short every day. Marketing campaigns often don't boost sales. R&D investments yield nothing. As with these conventional failures, there's a lot to learn from environmental missteps.

As vast as the opportunities for Eco-Advantage are, successfully exploiting them is not easy. To pretend otherwise is foolish. That's why we spell out below thirteen common reasons environmental initiatives fail, along with some thoughts on how to avoid these pitfalls.

1. SEEING THE TREES BUT NOT THE FOREST

Ford committed substantial sums of money to making the River Rouge factory green, but manufacturing is not the core environmental problem in the auto industry. Instead, the company should have studied its environmental footprint and taken a hard look at its products. While Ford focused on planting grass on the River Rouge roof, Toyota was rolling out its hybrid gas-electric technology and taking the auto world by storm. Instead of leading the charge, Ford now must play catch-up. It has licensed some of Toyota's hybrid technology and has committed to producing 250,000 gas-electric vehicles a year by 2010.

We don't mean to pick on the Ford Motor Company or on Bill Ford in particular. Missing the real issue is a common problem in corporate environmental strategy, and Bill Ford's enthusiasm for green causes is admirable. If every CEO had his dedication, the world would be a far better place. Our point, really, is that enthusiasm is not enough. Thoughtful choices translate commitment into the kind of meaningful action that gives a company an Eco-Advantage.

Consider Ford's $25-million grant to help create Conservation International's Center for Environmental Leadership in Business. It's a noble gesture, but the NGO's agenda centers mainly on promoting biodiversity, especially in tropical rain forests. Although a pressing environmental problem, biodiversity is not Ford's central environmental issue. Ford and its new partner looked for ways to work on issues more in line with Ford's environmental challenges, like climate change, but it never really came together. So *someone* will benefit from the money, but the company that gave it won't be at the front of the line.

Solution: Know Thyself

To avoid focusing on the wrong things, companies need to understand where they are environmentally vulnerable. Tools like Life Cycle Assessment and AUDIO analysis can help to bring the forest, not just the trees, into focus. Partnering with NGOs and experts to get outside perspectives on how the public perceives the company can also be very important. The basic principle is simple: Focus limited resources on the issues most central to the company's environmental footprint and reputation.

Compare, for example, Ford's odd choice of Conservation International with Wal-Mart's more recent partnership with the National Fish and Wildlife Foundation. The mega-retailer will spend $35 million to preserve 138,000 acres of wildlife habitat, equal to the area the company uses for stores, parking lots, and distribution centers. This project goes to the heart of the congestion and sprawl problems that Wal-Mart's critics focus on. That's a good fit.

Don't ignore the full value chain either. Remember how Unilever, merely a fish buyer, helped create the Marine Stewardship Council and has invested significant resources in time and money to promote sustainable fishing. Coca-Cola knows that it can't escape responsibility for what its bottlers do, even though they are entirely separate companies. So it has a VP for Environment and Water Resources in Atlanta helping to coordinate water conservation initiatives all around the world.

2. MISUNDERSTANDING THE MARKET

There's a fine line between being a step ahead of the pack and miscalculating the marketplace. Interface Flooring tried to promote sustainability by "servicizing" the commercial flooring business and renting carpets to its customers. The idea seemed to make sense: Interface would lay the carpet and take it back as the product wore out, facilitating recycling. But, as we described earlier, this eco-friendly new business model failed—the market just wasn't built for rented carpets. The company earned goodwill in the marketplace from this effort, but customers weren't interested in paying for flooring annually out of the operating budget instead of in one lump sum out of the capital budget.

Interface is not alone in having misjudged the marketplace's readiness for green products or services. Many big, successful companies have similarly tripped up. Unilever's attempt to source its fish sticks from sustainable fisheries has been slowed as consumers in some markets balked at the company's attempt to substitute a more plentiful white fish (called hoki) for cod. As frozen foods marketing executive Dierk Peters told us, "Some people call the UK 'Cod's own country.'

. . . cod is the gold standard, so there was real consumer reluctance to buying the hoki."

Remember how misunderstanding the difference between the U.S. and European markets brought biotech giant Monsanto to its knees in the 1990s. The company had great technology for genetically modified organisms, and CEO Robert Shapiro made sustainability the centerpiece of his business strategy. But the rollout failed miserably because Monsanto failed to pick up on a basic market reality: European consumers were deeply uncomfortable with the modified foods.

Solution: "Blocking and Tackling"

It's easy to get excited about a new environmental initiative. Eco-products hold up the promise of capturing new markets while making the world a better place. But the process for deciding whether a green initiative makes business sense should be fundamentally the same as for any other new product: What is the market need we're filling? Is there a customer base? What will our cost structure look like? Are others already filling this niche? Is our Eco-Advantage protected through patents or other proprietary information, or will new competitors easily enter the market? In short, the initiative might be green, but the usual blocking and tackling still needs to be done.

Be sure to think long-term as well as short-term. Consider intangible factors such as corporate image and reputation, regulatory burdens lifted (or taken on), and customer loyalty. To the extent possible, put hard numbers on intangibles. What is customer loyalty worth? Start with the cost of customer acquisition. What about employee morale? Start there with employee churn costs. Don't avoid the basic process of evaluating probabilities of risk and reward just because it's hard to put numbers on some benefits.

3. EXPECTING A PRICE PREMIUM

As we made clear in Chapter 5, selling products on greenness *alone* rarely works. Quality, price, and service remain critical to most customers. Patagonia sells green products that use organic cotton and recycled materials. Customers would not stay loyal, however, if product quality faltered.

Sometimes customers will pay a premium for environmental qualities or a green image. Patagonia sells at the high end of its market, and Toyota's Prius sells for $5,000 more than comparable cars. But these exceptions do not disprove the rule. In today's markets, green products cannot regularly command a higher price. From Shell's experience with eco-friendly Pura gasoline in the Netherlands, to the many attempts to make compact fluorescent bulbs palatable at up to ten times the cost of regular bulbs, green premiums are tough to pull off.

Even products that *save* the customer money over time, like energy-efficient appliances, often struggle to find a foothold. If they succeed, they are sold not on environmental attributes alone but on long-term cost-savings or durability. People simply do not like to pay more up front. As economists like to say, customers have an extremely high "discount rate" on purchases. They see money in their pockets now as more valuable than money saved in the future.

This fundamental customer truth makes the sale of green goods difficult. For example, Timberland has found it hard to sell organic cotton T-shirts, which cost 25 percent more to make, at a profit. The shirts offer customers no real functional improvement. The pitch is all emotional. So it's challenging to induce customers to pay a premium for an environmental feel-good moment. Patagonia has had some success charging more because it has built such extensive brand equity on its environmental leadership. But, frankly, it's not a public company and cares much less about earning a profit on every item.

Solution: Green as the Third Button

As we've stressed, don't pitch *only* the green attributes of a product. Look closely at Toyota Prius ads and see the dual message. The car has an award-winning power train that will give you some zip, and it's got cool technological features. It *also* helps to save the planet. The environment is the third "button" pressed. Even the greenest companies, like Patagonia, don't pitch only the environmental aspects of their products. They talk first about quality. And after years of delivering on that promise, Patagonia is positioned to define product quality to include an *environmental* dimension.

4. MISUNDERSTANDING CUSTOMERS

It's no surprise that McDonald's restaurants generate a lot of waste. But who would guess that over 30 percent of their waste, by weight, is liquid? When landfills charge by the pound, water is an annoying thing to pay for. Ten years ago, McDonald's Sweden started asking customers to pour ice and beverages into a separate bucket from other trash. It worked beautifully. Over 75 percent of customers willingly sorted their own garbage. Waste weight went down 25 percent, saving McDonald's millions of krona. But when McDonald's tried the same thing in the United States, consumers didn't go for it. Americans just aren't Swedes, and vice versa.

When an environmental initiative depends on changing customer behavior, be careful. If the change doesn't save people time or money—and sometimes even if it does—they may resist.

Solution: Know Your Customers' Limits

Every coffee drinker on the go knows the problem. To keep your hands from burning, one paper cup won't do the trick. Yet, giving everyone two cups seems wasteful. So Starbucks decided to tackle this conspicuous waste problem. The company designed a new cup with a built-in exterior insulating layer to keep fingers cool. The cup cost more, but it contained more recycled content and customers would only need one—a real win-win for the environment and bottom line . . . except tests soon showed that coffee drinkers still wanted two cups.

When executives ran the numbers, they determined that if more than 10 percent of customers took two cups (and Starbucks would not refuse a customer's request), the environmental effects would be worse than the current situation. Starbucks knew that trying to reform customer behavior can be difficult. So the company found a middle ground, the now ubiquitous coffee-cup sleeve. The cardboard piece slides over the cup and protects customers but uses 40 percent less paper than a full second cup.

Some environmentalists would call this story a failure, but it's not. Rolling out the environmentally preferable cup would have failed on financial *and* environmental grounds when customers took two. Starbucks made a hard decision to abandon an innovative solution, found a middle ground, and avoided failure. As Sue Mecklenburg, Starbucks' VP of Business Practices, told us, expecting customers to make the right environmental choice often leads to disappointment.

5. MIDDLE-MANAGEMENT SQUEEZE

More than anywhere else in the organization, middle management is where the rubber meets the road in the drive for Eco-Advantage. It's also where environmental efforts often break down. The senior executive team hears the CEO's call for an environmental focus, and line workers often welcome the chance to make their companies more eco-friendly. But middle managers are pulled in many directions. They face critical trade-offs and hard choices on a day-to-day basis. They're told to increase sales and throughput, cut costs, fatten profit margins—and now, be green as well.

Incentives often aren't aligned with green goals. End-of-year performance reviews generally turn on the company's core concerns, not environmental targets. BP Senior Advisor Chris Mottershead describes this issue as "a tension between business performance and environmental goals."

Mottershead's boss, CEO Lord John Browne, committed BP to reducing its climate footprint, which meant cutting greenhouse gas emissions from the company's refineries. The easy way to reach the goal was to cut throughput. But plant managers had production goals that conflicted with the greenhouse-gas reduction effort. "The stumble was that we told refineries to chase volume *and* produce cleaner fuels *and* lower greenhouse gases," Mottershead told us. "Those three goals are *not* independent—you drive for one and it has real consequences for the others."

We call this problem the Middle-Management Squeeze.

It's very common, especially among companies making bold environmental commitments. CEO leadership and stretch goals are still success factors, but they can also cause irreconcilable conflicts for the people caught in between—the ones actually running the company's operations and trying to meet sales goals, keep costs down, and hit profit objectives. Layering environmental concerns onto that mix can lead to overload.

***Solution:* Incentives and Training**

Ignoring the Middle-Management Squeeze is a recipe for trouble. So deal directly with the multiple pressures on middle managers.

Executives at this level almost always follow their incentives. If they don't get signals that environmental success is part of their job, they won't prioritize it. Writing environmental goals into a manager's job description is one way to ensure focus. Hard cash doesn't hurt either—so put environmental goals into bonuses if cultural pressure isn't enough. Creating environmental metrics and making them part of the company's key performance indicators is also helpful. And hold people directly and publicly accountable for their group's results—remember GE's Session E review.

Guidance from above must also be part of the package. At BP, the physics of refining wouldn't relent. So the company leaders did the only thing they could—they told the refineries to operate as efficiently as possible, but to stop worrying about *total* greenhouse gas emissions. The refineries' job was to maximize economies of production and lower *per unit* greenhouse gas emissions. Reductions, the executives knew, would have to come from elsewhere in the system.

Training is also a vital tool. As part of a seventeen-day Leadership Development Program, 3M's environmental executives teach managers about some of the trade-offs they will face and pose some questions: What should you do when the lowest-cost supplier seems to have no environmental standards? What do you do if you can either invest in a product line or fix an environmental problem? Like middle-management everywhere, 3M plant managers deal with the conflicts as best they can. They've internalized the fact that they somehow have to hit cost, quality, and environmental goals simultaneously.

6. SILO THINKING

Classical mythology tells the story of how Hercules tried to kill the multi-headed monster Hydra. Every time he smashed one head with his club, two would grow back in its place.

Environmental executives trying to combat pollution and other environmental problems can sometimes feel like poor Hercules. Invest in a scrubber to reduce air pollution and you've got a waste disposal problem with the captured sludge. Store the sludge outside and you may develop a water pollution issue.

"Our failures are usually when abatement goes wrong," Intel's Tim Mohin told us. "Waste treatment solves one problem and creates four others." For years, all semiconductor companies have struggled with ways to reduce use of perfluorocarbons (PFCs). Used in the etching step of chip production, PFCs are up to 10,000 times more powerful than carbon dioxide as a greenhouse gas. Early attempts to reduce PFCs focused on capturing the gases after they were used. But a giant abatement system at Intel actually increased energy use and greenhouse gas emissions, and its recycling system created other airborne toxics. "The side effects were worse," said Mohin.

Both Hercules and the managers at Intel were trapped in what we call silo thinking. Instead of searching the whole barnyard for answers, they were focused on one narrow part of the operation. The solution didn't fully solve the primary problem, and it created new ones.

Silo thinking can also lead to missed upside opportunities. Take Fuji Photo Film's experience with disposable cameras. Another firm, Jazz Photo, built a profitable business refurbishing discarded Fuji disposable cameras. Jazz sold tens of millions of recycled Fuji cameras before Fuji took it to court for patent infringement. But here's the environmental strategy question: Why did Fuji miss this market opportunity, leaving hundreds of millions of dollars on the table? Part of the answer is silo thinking. Fuji made good disposable cameras. It likely saw the used cameras as a waste product rather than as an input to a second product, a recycled disposable camera.

Solution: Design for the Environment and Value Chain Thinking

Hercules solved his dilemma with the Hydra by looking at it in a different way. Instead of smashing heads that doubled each time he destroyed them, he had his nephew cauterize the neck stumps before the monster could grow new ones, thereby stopping the problem at its source.

Intel's solution was a bit less dramatic. It established a Design for the Environment program. Twenty environmental professionals now work side-by-side with product and process designers, as well as the basic research scientists who explore ideas for production six or more years out. Together, they systematically identify and design out environmental problems before they crop up. Toxic emissions are still a big burden. But they are down, and other environmental problems have been avoided entirely by life-cycle thinking.

If pollution is a cancer on the earth, abatement is chemotherapy—a treatment after the problem exists. Design for the Environment is like quitting smoking and eating right. It's preventive medicine for environmental problems. It can't eliminate every risk, but incorporating environmental considerations into product design goes a long way.

7. ECO-ISOLATION

To make Eco-Advantage an everyday part of business strategy, companies need champions who are passionate and knowledgeable. But relying solely on a designated environmental team creates its own problems. As one executive put it, "Just saying, 'throw it to the green guy' doesn't work." We see three related problems that stem from isolation.

For starters, some worthy initiatives may die on the vine. FedEx Kinko's has a real champion in Larry Rogero. For years, he's been on the environmental case at the copy giant, taking on whatever role is needed, including "instigator, motivator, accountant, and sometimes implementer." For a long time, he was also a one-man band. True, FedEx Kinko's has achieved a great deal, including significant renewable energy purchases and large-scale use of recycled-content paper. But the company is really just starting to bring environmental thinking deep into the core of the business.

Some interesting initiatives, like the self-serve "green machines" that let customers make copies on 100 percent recycled paper, lost steam because they had limited support outside Rogero's department. After the test promotional period was over, individual branch managers assumed responsibility for offering the eco-product. Those that saw a real benefit for employee and customer loyalty continued the program. But most didn't.

Why should a worthy eco-idea become nothing more than a short-term promotion? In part, the answer is eco-isolation. Because the idea bubbled up from an isolated environmental department, the organization didn't provide the broad support needed to determine whether there were lasting benefits—in reputation, sales, or customer loyalty—that might argue for making the green machine a permanent change. Leaving the final decision up to individual line managers didn't help either. In the absence of compelling incentives or goals, why would the managers choose an option that might cost them more upfront? This conflict is the essence of the Middle-Management Squeeze we discussed earlier.

A second aspect of eco-isolation is overly tight budgets. We've seen big companies try to handle environmental issues on a budget so small that the environmental managers can't get anything done. It's easy to underestimate what it takes to get these issues right. Eco-Advantage does not emerge out of thin air. Environmental strategies require good data collection and analysis, just like any other strategy. Remember: Mishandled environmental issues can cost a company a great deal—in hard cash, reputation, customers, or employee morale. Understaffing or underfunding the environmental group won't generate Eco-Advantage. Indeed, doing the work on the cheap may cost far more than it saves.

Finally, eco-isolation can lead to the classic problem of "one hand doesn't know what the other is doing." Toyota, for example, has joined in with other auto companies to fight regulations that raise fuel efficiency standards, even suing California over their new laws. We're sure the company's executives are aware of this action. But do the government relations people *really* know that the lobbying efforts threaten to undercut all the good work that has been done on positioning Toyota as an environmental leader? Frankly, given Toyota's market-leading position on fuel efficiency, why wouldn't the company argue for more stringent standards?

Companies are often pulled in two directions. Sometimes they purposely say one thing to one audience, and something quite different to another. But it's also possible that these decisions are being made by two different groups that don't communicate that often. Or perhaps the full strategy hasn't been laid out for everyone. Eco-product design teams and the marketing staff are not often in the same room with government and regulatory affairs people. That isolation can come at a high price.

Solution: Top-Level Commitment and Integration

Solo environmental champions almost always fail. Successful corporate environmental strategies build on thinking from across the company. Integrating the concerns, needs, and incentives of those on the company's operational front lines into the game plan is essential. Environmental managers can guide the process, but line managers must own the initiatives. Even the best-conceived Eco-Advantage strategy will ultimately fizzle if it is not backed by CEO commitment and connected with the managers and workers who must implement it.

FedEx Kinko's is working on tying its environmental vision more closely to operations. With CEO support, Rogero took the company's top executives through sustainability training. He also launched a mission-setting process with a focus on environmental issues called "Six Flags on a Hill." One executive told him, "It makes you think that this is about a lot more than just recycling, but I've had enough science. . . . what can we *do*?" This is precisely the reaction you want: engagement and eagerness to get going.

FedEx Kinko's created teams to focus on a core set of issues, such as energy and community involvement, and gave leadership roles to a spectrum of executives. One team explored how the company could help commercial customers become more sustainable. They found that customers were excited at the prospect of letting FedEx Kinko's do more of their document management—cutting costs and reducing paper use. The "Six Flags on a Hill" team got more engaged in the pursuit of eco-initiatives when they saw the potential to drive customer value and expand sales.

Showing executives how to look at issues through an environment lens can spur fresh thinking, but the larger goal has to be breaking down the barriers between environmental and business strategy. Efforts to promote cross-fertilization of line and environmental managers and cross-cutting initiatives, such as those we've observed at IKEA, 3M, and DuPont, all pay dividends. They tell the organization that environment is a central part of business strategy.

8. CLAIMS OUTPACING ACTIONS

In the eagerness to be green, sometimes companies make promises before they've taken any action. NGOs will quickly jump in and cry "greenwashing," but that's overly simplistic. Real greenwashing is when a company claims to be doing something green and knows full well it isn't. The failure we're talking about here is one of action, not of intent.

Commitments to reduce the environmental impact of your product line—before those products are even close to ready—fall into this category. Often it's another failure of eco-isolation, where the marketing folks get ahead of the curve and nobody from the environmental side is close enough to the action to rein them in. It's not an unusual problem in companies. Execution often lags behind intent, but the stakes on environmental claims are especially high.

Solution: Do What You Say and Know Your Stuff

The most straightforward—and blindingly obvious—way to avoid getting ahead of yourself is to make everything you say true. Public promises must be rooted in real design and process changes that generate actual improvements in environmental outcomes. Ignorance about environmental impacts won't fly. More and more, companies are expected to have hard data to prove their claims.

The moral: Tread carefully and make claims you can meet. Put in place tools that help you check them, too, like 3M's Marketing Claims Committee. Don't think small or set your sights low. That's admitting defeat at the front end. Market-changing innovation in products or processes remains a basic way to generate Eco-Advantage. And don't stop setting stretch goals and living by the Apollo 13 Principle. With important issues like these, a company shouldn't take no for an answer. But when you go public, make sure you know what you're talking about.

9. SURPRISES: WASP STINGS AND UNINTENDED CONSEQUENCES

Herman Miller's facility in Holland, Michigan, called the Greenhouse, is one of the most environmentally sound manufacturing sites in the world. The building is efficient, productive, and beautiful. And though you wouldn't think a factory could be calm and relaxing, the Greenhouse pulls it off. The exterior reflects environmental sensitivities as well, with landscaping that keeps wetlands intact and leaves local western Michigan plants and grasses in place.

But nature presents challenges. Soon after the green manufacturing site opened, employees complained of wasps in the parking lot. It turned out that the fields of wildflowers surrounding the building were attracting the pests. It's a small story, but a telling one. Making an environmental choice—in this case keeping natural landscape intact around the green building—can take a strange turn. This story has a happy, even environmentally sound, ending: The company realized that bringing in bees to pollinate flowers would enhance the look of the fields, drive off the wasps, and reduce the total pest count.

Value chain and life-cycle effects can often be surprising. A new system to save water uses more energy. A product redesign to eliminate a dangerous chemical changes the performance of the product in unforeseen ways. Don't be surprised. Expect the unexpected. And when you come up against unintended consequences, see if you can turn them into opportunities.

Solution: Walk Before You Run

Although environmental thinking can lead to sustained and extensive competitive advantage, it's helpful to start with modest expectations. To maximize the odds of success in executing environmental strategies, keep a few key points in mind:

First, walk before you run. Start with pilot initiatives. Unilever launched its commitment to sustainable agriculture by operating test farms on a few pilot crops to experiment with low-water and low-pesticide farming methods.

Second, take a systemic view. Analyze in advance the life-cycle consequences of a new initiative. When you pull a lever here, understand what happens elsewhere in the value chain. Will saving energy within your walls somehow mean more energy used by customers?

Third, be careful about forecasting potential gains. Unilever's eco-friendly farms are generating good results. But they vary across crops and geographies. So the company is not claiming that all the benefits will be easily rolled out in all crops worldwide.

10. PERFECT IS THE ENEMY OF GOOD

Here's a tale of two companies, The Body Shop and McDonald's. The Body Shop set out from its founding to be a different kind of place. As its website will proudly tell you, the company is against animal testing. It supports community trade, seeks to "activate self-esteem," defends human rights, and protects our planet. The Body Shop published its first environmental statement, "The Green Book," way back in 1992 when such ardent and open commitment to environmental issues was very rare in the corporate world. It phased PVC plastic out of its products years ahead of the competition.

No doubt about it, The Body Shop is a green company. Until recently, however, it was not a gold one. Pursuit of its environmental and social mission in a manner that was inattentive to economic realities kept the company from consistently turning a profit. That's changing, but getting there has been a long, hard road.

In a very fundamental way, sustainability depends on long-term economic success. It's the only way to fund whatever degree of environmental commitment a company chooses to make.

McDonald's can claim some deep green roots, too. The company has been working to reduce the impact of its packaging for over fifteen years, constantly looking for ways to reduce its environmental footprint. But McDonald's is also a company that keeps economic realities squarely in mind.

A case in point: A few years ago, in Europe, the company tried three new options for its McNuggets package: a polystyrene container, a cardboard box, and a paper bag. It considered the options against cost, environmental impact, and the three main criteria of user experience, function, look, and feel. The polystyrene was solid on user experience (it kept the chicken warm) and it was relatively cheap, but it was the worst environmental choice and thus was unacceptable to McDonald's customers in many countries. The paper bag was the best option for the environment, but customers didn't like the experience—the McNuggets cooled quickly and the paper felt flimsy. The middle option, the box, satisfied customer needs but used a lot more material and took a greater toll on the environment.

Confronted with no perfect answers, McDonald's could have continued to search for the Holy Grail of low-cost, high-functioning, eco-friendly packaging. Instead, the McDonald's packaging team chose to settle for something less than perfect but better than what they had already. The company adopted a paper-based clamshell that worked as well as the box but used 30 percent less material (see figure).

McNuggets Packaging Options: Performance against Key Criteria

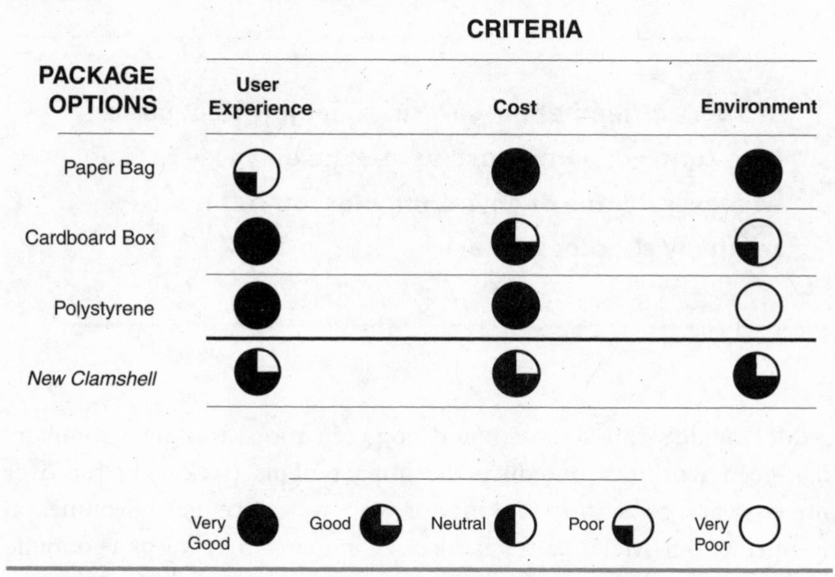

A perfect solution? Far from it. The clamshell used more material and cost more than the paper bag. Confronted with the same range of choices, perhaps the old Body Shop would have kept on looking—or gone with the greenest solution, the bag. But sometimes close is good enough. No business can be sustainable if it pursues environmental purity without regard to business consequences.

Solution: Prepare for and Accept Trade-offs

Flawless solutions are few and far between. WaveRiders know this. They don't search solely for win-win answers because they know that the quest for perfection too often leads to inaction. Like McDonald's, successful companies often find incremental solutions that steadily improve environmental performance, while minimizing the burden in other areas. Some progress is much better than no progress.

"There are messy trade-offs in *anything*," a Herman Miller executive told us. It's a corollary to the "wasp stings" problem: Trade-offs are the norm, not the exception. No company can jump on every opportunity and sometimes the costs of going green are just too high. But often partial wins are possible. It helps to look broadly at the costs and benefits of any strategy decision—upstream and downstream impacts, short- and long-run implications, concrete gains or losses, and intangible effects. Sometimes new options emerge. As Timberland's Terry Kellogg told us, "Often the trade-off is not business win vs. not-business win, but short-term vs. long-term business win."

This "take what you can get" strategy may seem to fly in the face of the Apollo 13 Principle ("no is not an option") we discussed earlier. But it's not so. McDonald's did *not* accept that environmentally sound packaging would cost too much or diminish the customer experience. In that sense, its leadership team *did* follow the Apollo 13 Principle. They accepted some cost increases to achieve the brand quality and environmental goals—a good, but not perfect, outcome.

IS ROHNER A FAILURE?

Our research suggests that Swiss fabric manufacturer Rohner Textil would rank near the top of any list of sustainable companies. Its products are nontoxic and biodegradable, and its manufacturing is state-of-the-art and low in every kind of pollution—water, air, even noise. But in the ten years since Rohner produced the most sustainable textile in the world, the company has not become a titanic success. Rohner has not dominated the textile marketplace. Toxic dyes are still the norm in the fabric industry. In fact, Rohner has priced itself out of most markets, almost ensuring it will stay small. So is Rohner a failure?

The answer lies in the eye of the beholder. Rohner broke new ground but found only a limited market for its eco-friendly fabrics. It has not grown substantially. Yet Rohner *is* profitable, and it provides well for its customers, employees, community, and the Earth.

11. INERTIA

Corporate culture and traditional ways of doing things in a company help to ensure consistency, excellence, attention to quality, and other virtues. But these same positive attributes can lead to inertia and make it hard to incorporate environmental concerns into strategy.

As Stephan Schmidheiny, GrupoNueva founder and former head of the World Business Council on Sustainable Development, tells it,

> After the 1992 Earth Summit . . . I was excited about putting my big new company on a sustainable, eco-efficiency footing, so I explained it all at a meeting of my key executives. They were supportive, excited; they came up afterwards and told me so. Of course, nothing happened. The strongest force on Earth is inertia, defined by some cynics not in physics terms but as "people resistance to change."

Environmental initiatives often ask people to step outside their comfort zone. Maybe designers have never been asked to minimize environmental impact before, or engineers have never been asked to optimize for waste reduction, not cost control. Moving people into a new way of thinking won't happen without a healthy push.

Solution: Vision, Both Big and Small

Overcoming inertia comes in two steps. We'll let Schmidheiny continue: "It was not my people's fault; I had done absolutely nothing a leader must do to cause change except to describe a vision—and this is only the first step. . . . Then the leader must break the vision into 'bite-sized chunks' of objectives, action plans, and measurable results."

The vision is critical. Where does the company want to be in one year? In ten? What environmental goals should the company shoot for? But every sweeping vision needs to be brought down into actionable steps. If the goal is to cut greenhouse gas emissions 30 percent, what needs to happen first? Perhaps it's a life-cycle analysis on core products or process mapping to find out where the emissions happen. In Chapter 11 we lay out one plan of attack in the short-, medium-, and long-term for developing broad, effective environmental strategies.

12. IGNORING STAKEHOLDERS

In 1968, geologists found a gold mine (an actual one, along with copper and silver) in Flambeau, Wisconsin. A subsidiary of Rio Tinto jumped on the opportunity and got to work digging up the valuable metals . . . in 1993. What happened during that missing quarter of a century? Well, the company failed to get the permits it needed. It badly mismanaged relationships with regulators, the community, and other stakeholders.

The original site plan was standard fare for the 1970s: an open pit mine with a holding area for toxic waste (called tailings). After the mine was played out, Rio Tinto would flood the open pit and create a lake. As Rio Tinto executive Dave Richards retells the story, "Local community reaction was negative in almost every respect." Dug-in opponents can delay a project like this for a long, long time. Finally, the company changed the plans to include a range of more sensitive actions. It would shrink the acreage of the site, rebuild the mine afterward to create recreation areas, and monitor the vegetation and groundwater for forty years. Digging started in the 1990s, and the site recovery plan is in effect today.

This may seem blindingly obvious by now, but we'll say it again. Stakeholders matter. Rio Tinto ran into some familiar problems. As we discussed earlier, Alcan misunderstood native and environmental community opposition and stranded $500 million in sunk costs on a half-built tunnel. Shell's well-known experience trying to dispose of the Brent Spar oil platform represents an even starker case.

Companies, especially multinationals, need permission from society, employees, regulators, and many others to grow—even to exist. Failing to get buy-in—or what we could call "letting the license to operate expire"—can cause problems with a range of stakeholder groups. Starbucks had problems when it rolled out a "preferred supplier" program to encourage conservation on coffee plantations. The first attempt had unclear guidelines and didn't address small farmers well. After some helpful feedback from a range of stakeholders, Starbucks developed flexible guidelines and best practices, all laid out in a hundred-page manual.

Companies can't just go about their business or launch new initiatives without buy-in all along the value chain, and even outside it.

This is a networked, interdependent world. Going it alone is a doomed strategy.

Solution: Map the Stakeholders and Get Them Involved

Knowing your stakeholders is essential. Start by mapping out which groups focus on particular issues, as we discussed in Chapter 3. And build relationships with NGOs and others so that you're not trying to open a dialogue when a crisis is at hand. Don't be afraid to sit down with even the harshest critics. It will pay dividends to build connections and develop understanding of their concerns and priorities.

The bottom line is simple enough: Listen to outside perspectives. We must offer two quick caveats. First, you can never satisfy every stakeholder interest. Sometimes NGO concerns are misplaced. So don't be shy about pushing back a bit. When Greenpeace leaned on shoemakers to reduce PVC use, Timberland questioned the level of pressure. The company was starting to phase out PVC in its shoes already, not to mention that the shoe industry was a very small player in the PVC problem. So Timberland refused to be intimidated. Why spend all your energy on the shoe companies, company officials asked Greenpeace, when the construction industry uses dramatically more PVC? Shouldn't the campaign focus at least some of its attention there?

From Greenpeace's point of view, shoes have big, customer-facing brands, and the construction industry does not. As Timberland's Terry Kellogg recalls, "They came to our industry because they hoped to get traction on the issue . . . and it worked." But companies should push back if they have good reason. Of course, NGOs have extensive knowledge, but if you know your life-cycle issues better than they do—and you certainly should—they might listen. Address the *real* environmental problems, and most of the NGO community will get it, appreciate it, and potentially make a big public display of support.

A second caveat: Don't focus so much on stakeholder feelings that you forget to address real problems. It's not all about buy-in and communication. Shell has gotten so enamored with stakeholder relations that it can sometimes fall prey to this problem. After Shell's

SAPREF crude oil refinery in South Africa "significantly" underreported air emissions, the company focused initially on the community and communications failures. These issues were real and important, but what about the actual emissions problems? Yes, feelings are facts, but sometimes facts are facts.

HEAPING FAILURE ON FAILURE

Some companies seem to have missed the lesson from the famous Johnson & Johnson Tylenol scare in the 1980s. When seven people died from poisoned pills, the company immediately recalled millions of bottles, took responsibility, and refused to put Tylenol back on the shelves until it had developed tamper-proof packaging. In comparison, ExxonMobil has been battling in the courts over damages from the *Valdez* oil spill for almost two decades. Likewise, when a tanker run by French oil company Total sank in 1999, the company mismanaged public perception badly. As an industry trade magazine said, "In the immediate aftermath, it seemed as if TotalFinaElf did not feel it was responsible. Six months after the accident took place, Total put out television adverts trying to tell people what they were doing. They went down badly. . . . The damage to public opinion within France had already been done." Waiting too long to say *"mea culpa,"* even when it's unclear who's at fault, can be much worse than being right.

13. FAILING TO TELL THE STORY

Recently, a well-known company made a big public announcement about a new environmental policy—including a bold assertion that henceforth all employees would make decisions differently. But when we e-mailed a friend at the company and asked her how her job would be affected, she said, "This is the first I heard of it." Oops.

Is forgetting to tell the story internally a failure? You bet. How about missing the chance to tout your good deeds in general? Yes, a minor one, but it is a lost opportunity. Like the proverbial tree falling in the forest, does an initiative accomplish much if nobody knows about it? Eco-Advantage comes from taking action—and getting credit for it.

Solution: Storytelling and Green Marketing

Companies need to tell their own employees what they're doing. Educated employees find even more opportunities for Eco-Advantage, and they inspire customers. Green marketing to the outside world also has its place and is a vital tool in the arsenal. If the actions are significant and the environmental benefits justified, then let everyone inside and outside the gates know about it.

Many companies tell us that they don't want to say something if it brings too many questions. Remember Levi's silent switch to 2 percent organic cotton, with no marketing? The company was concerned that people would have questions about the other 98 percent. In some cases, this concern is justified, and we've cautioned many times to avoid public statements that you can't back up. But companies need to leverage clear wins. That's how you build morale and momentum toward sustained Eco-Advantage.

ECO-ADVANTAGE BOTTOM LINE

No set of strategies can guarantee success. Many environmental initiatives, no matter how well designed and focused, will fail. But looking out for the pitfalls highlighted here will maximize the odds of success. Below, a quick review of the 13 Failures and the recommended solutions from the Eco-Advantage Toolkit.

Failure	Solutions and Tools
1. Seeing the trees but not the forest	— Know your own issues (AUDIO, LCA) — Data and metrics — Partnerships and outside perspectives
2. Misunderstanding the market	— "Blocking and tackling"
3. Expecting a price premium	— Green as the third button
4. Misunderstanding customers	— Know customer limits and drivers
5. Middle-management squeeze	— CEO commitment and guidance — Incentives — Engagement and training
6. Silo thinking	— Value chain thinking — Life Cycle Assessment — Design for the Environment
7. Eco-isolation	— Broad-based executive commitment — Ownership at the operational level — Cross-fertilization between environmental and line managers
8. Claims outpacing actions	— Data and verification — Internal vs. external goals
9. Surprises: Wasp stings and unintended consequences	— Value chain thinking — Pilot programs — Conservative estimates of wins — A sense of humor
10. Perfect is the enemy of good	— Anticipating and accepting tradeoffs — Extended perspectives
11. Inertia	— Vision — Execution in bite-sized chunks
12. Ignoring stakeholders	— Stakeholder maps — Partnerships — Know that feelings are facts
13. Failing to tell the story	— Storytelling, both external and internal — Training

Chapter 11 Taking Action

The plays, the mindset, even the tools for building Eco-Advantage mean little until they are translated into action. Many companies have pieces of an environmental strategy in place, but few are systematic about driving environmental thinking into their approach to business.

Because no two businesses are the same, every company has to plot its own Eco-Advantage path. We've found, however, that the pace of progress is greater if certain things get done first. In this chapter, we spell out an implementation plan and suggest a program of action. Broadly speaking, the agenda is:

- Short term: Find out where you stand and launch pilot projects.
- Medium term: Track performance and build an Eco-Advantage culture.
- Long term: Drive environmental thinking deep into the business strategy.

SHORT-TERM ACTIONS: WHICH BALLS ARE IN THE AIR?

Lifting corporate environmental strategy to a higher plane begins with clarifying where the company stands and what needs to be done.

The Big Issues

Business executives today must handle a wide range of competing concerns. Globalization, the spread of the Internet, outsourcing, cost pressures, competition from everywhere and anywhere—the list is getting longer and executives' tenures shorter. In the environmental realm alone, the number of balls in play increases every day. In Chapters 2 and 3, we outlined ten big environmental problems and twenty players that can dramatically affect a company's fortunes. Both lists are expanding, not contracting. Without some attempt to understand the issues and players, managers can quickly feel overwhelmed. They know they're juggling a lot, but they aren't sure which issues are just balls and which are flaming chainsaws.

The short-term agenda, then, is to discover where to focus attention. To get started, we recommend undertaking three kinds of analysis centered on:

- the environmental issues that touch the business
- what stakeholders think about the environmental performance of the company
- whether the company has the capabilities it needs to address its environmental challenges

1. AUDIO ANALYSIS: ISSUE SPOTTING

To figure out which way to go, you have to know where you are. Let's go back to the AUDIO tool we discussed in Chapter 2. This grid has major environmental issues on one axis and five dimensions to explore on the other—Aspects, Upstream, Downstream, Issues, and Opportunities. AUDIO helps the company "listen" to the business and understand the concerns that must be managed up and down the value chain. Like a traditional "SWOT" analysis (strength, weaknesses, opportunities, and threats), AUDIO is a tool for spotting downside risks and upside opportunities. But the addition of the

AUDIO Framework

Challenge	Aspects	Up	Down	Issues	Opps
1. Climate Change					
2. Energy					
3. Water					
4. Biodiversity					
5. Chemicals, Toxins, and Heavy Metals					
6. Air Pollution					
7. Waste Management					
8. Ozone Layer Depletion					
9. Oceans and Fisheries					
10. Deforestation					
11. Other Issues (industry-specific)					

value chain perspective represents a significant refinement on that basic tool.

To conduct the analysis, start quick and dirty. Begin by bringing a small group from across the company together and sketching a picture of the company's portfolio of environmental problems. Start with internal people who have environmental responsibilities, but include representatives from operations, design, marketing, purchasing, and customer service. Try, in just a couple of hours, to fill out the AUDIO matrix. Don't worry about the details or getting it all right. Educated guesses are fine.

First, companies should ask themselves what **Aspects** of these issues affect their own operations. Do we produce greenhouse gases? Use a lot of energy? Pollute the air or water? Use a species of plant or animal in our products that might be threatened? Consume a great deal of land? Then, getting more granular: Which particular product lines have toxic elements? Do some divisions produce more of these effects than others? And so on.

Next, ask the same questions about each environmental problem looking **Upstream** from your own operations. Do our suppliers use a great deal of water? What primary materials are we ultimately sourcing from sub-suppliers, and which issues are pressing on them? Look as far upstream as you can. Then look **Downstream** and ask questions about customer use and the end of the product's life. Is our product an energy hog? Does it pollute? What happens to our products when customers are done with them?

The next step is to look for vulnerabilities you need to address. Walk back through all of the items highlighted under Aspects, Upstream, and Downstream, and ask which elements create particular **Issues** or challenges for the company. Do some facilities or products depend on a steady supply of water? How will droughts affect those businesses or a supplier's business? Are some lines of business highly energy dependent? What would happen to costs if U.S. greenhouse gas emissions were legally capped and the government imposed a carbon tax or trading system? Do we use hazardous substances in our products or production processes? What would happen if a state or federal government banned these substances? Asking yourself these types of questions protects you against surprises—forewarned is forearmed.

Finally, brainstorm about the **Opportunities** that these pressures give rise to. Are you the most energy efficient producer in the market? Or would controls on greenhouse gas emissions create opportunities for you? Can you help customers deal with the issues *they* face? Remember that every problem or challenge presents opportunities for someone. DuPont made money selling the substitutes after CFCs were banned for damaging the ozone layer. Champion Paper supported spotted owl protection, knowing that its competitors had far more exposure to lands being taken out of timber production than it did.

AUDIO is not meant to be a one-time tool. Companies need to regularly reexamine the environmental balls in the air. Issues evolve, and as they do, you need to rethink the implications for your business. Every time you go through the exercise, you're likely to pick up new issues or see new twists on old ones. In some areas, AUDIO will create more questions than answers. When that happens, bring in concrete tools such as the Life Cycle Assessment (LCA). Looking at key products along their value chain can highlight new concerns or confirm the topline thinking from the AUDIO analysis. Either way, you'll be wiser.

At their heart, AUDIO, LCA, and all the "lay of the land" tools help to establish the Eco-Advantage Mindset. They position managers to look at their work through an environmental lens and create sensitivity to both challenges and opportunities. In a world of extended producer responsibility, they break a company out of its box and focus attention up and down the value chain, where environmental problems and opportunities can start. Use them to stretch thinking over time, across boundaries, and beyond the usual set of payoffs.

2. STAKEHOLDER MAPPING: WHAT DO OUTSIDERS THINK ARE THE KEY ENVIRONMENTAL ISSUES FOR THE BUSINESS?

It's great to know the facts, but as we've shown, feelings matter, too. You can do all the AUDIO and LCA analyses you want and still get blindsided if you're not in touch with stakeholder concerns.

Who are the key players you face? Figuring that out is a critical first step. So we suggest creating a stakeholder map that will track:

- **Rule-Makers and Watchdogs**: NGOs, plaintiff's bar, regulators, and politicians.
- **Idea Generators and Opinion Leaders**: Media, think tanks, and academic institutions.
- **Business Partners and Competitors**: Industry associations, B2B buyers, competitors, and suppliers.
- **Consumers and Community**: CEO and executive peers, consumers, "the future" (kids), communities, and employees.
- **Investors and Risk Assessors**: Shareholders, analysts and the capital markets, insurers, and banks.

Start by listing all the players you deal with now—or might want to connect with in the future—who fall into these categories. But think hard about groups or individuals that might be flying under your radar. Next, prioritize relationships across this playing field. Which groups have the biggest impact today? Which might emerge in the future? Follow that up with a few key questions: How much effort do we put into understanding the priority groups? Are there key individuals we need to get to know? Are we prepared for any concerns they might have?

Be sure to watch out for the classic battle between the urgent and important. It's easy to get swayed toward the squeaky wheels. But in focusing on the urgent, you might miss some stakeholders who could torpedo you down the road if you're unprepared. For example, banks may not seem vital . . . right up until the moment they decline to fund a project because of environmental concerns or liabilities. Likewise, beware of inertia. Just because you've had a relationship with a particular environmental group doesn't mean they are the right NGO to work with.

Players Influence Map

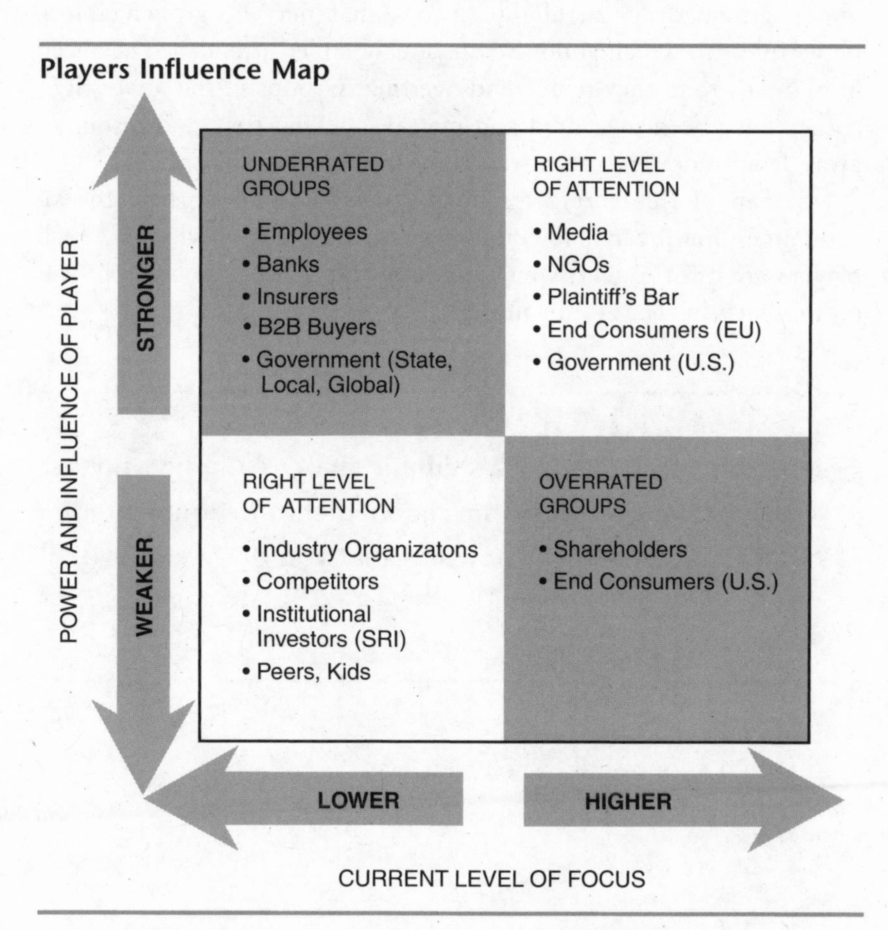

To help you avoid these kinds of problems, we've developed a tool to help prioritize the stakeholder world (see figure). In this example, we've taken a cut at mapping the players for a hypothetical business. First, array the players on two dimensions: (1) the power and influence they wield on your company and industry, from weak to strong, and (2) how much investment you're making in understanding them and managing the relationship—the worry level, if you will, from low to high.

Then group the players into four categories. In the upper right are the players deserving focus. They're influential and they're getting attention. The lower left shows the opposite: They're not very powerful, and they're not getting much attention. These two groups are being handled correctly. The other two groups are important. The upper left quadrant highlights groups that may be growing more powerful or to which you have previously under-attended. They need more focus than they're currently getting. Groups in the lower right corner have been overrated and may be sucking time and resources away from more important relationships.

No map of this sort perfectly portrays the playing field for all companies, but it can help you to develop a big-picture view of which players are most urgent, which are important, and which can be put on the back burner . . . for now.

The short-term focus is systematic analysis. Clarify who and what you've got to think about in the environmental realm. Gap analysis—knowing what you don't know—can be a big first step.

3. CORE CAPABILITIES ASSESSMENT: WHAT CAN WE DO TO IMPROVE THE ENVIRONMENTAL PROFILE OF THE COMPANY? WHAT ARE WE BEST POSITIONED TO HANDLE?

How does a company keep from getting bogged down when multiple environmental issues are vying for its attention? By focusing on the issues that matter most—and then determining what capabilities the company can bring to bear. For example, building on its technological strength, Toyota has emphasized the design of fuel-efficient cars. Moreover, Toyota has prioritized being the leanest manufacturer in the world. Pursuing excellence in the "Toyota Way" produces eco-efficiencies as a matter of course, without launching any special green manufacturing initiative.

We're not saying that companies shouldn't tackle issues beyond their normal scope. But they should do it in a smart way that leverages their talents. If the issue is tangential but still matters, look for partners. All the members of the Paper Working Group buy a great deal of paper. For some, like Time Inc. or Staples, paper policies are a real business risk. Working with other companies, they have a shot at changing the market. For other group members, such as Bank of America or HP, paper is an important but secondary issue to the core business. For those companies, partnering with others saves enormous time and resources.

The bottom line: Know yourself—your strengths, your weaknesses, and your resources for change. Then match that self-assessment with the environmental issues most pressing on your business. Only then will you know where you can score the most points.

First Actions

After the basic analyses have been completed, visible actions must follow quickly. It's important to put some quick points on the board to build momentum. We recommend three steps to focus everyone on successful execution.

1. CEO STATEMENT

Companies with no history of environmental thinking—and even those with some track record—need a statement from the boss committing the company to environmental values and goals. Upfront commitment plants the seeds of an Eco-Advantage culture, which the medium- and long-term action items will harvest.

Pushing the message out across the company is also important. This process often starts with a CEO conversation with top managers. Successful rollout then generally requires a broader statement to all employees, laying out an environmental vision and listing specific goals. A public statement should wait until real action is underway in the medium- or long-term. When companies get this backwards, they can look pretty dumb.

2. PRIORITY ACTION PLAN

If you've followed our action plan so far, you should know what environmental issues most touch your company, what outsiders are thinking about your eco-profile, and what your internal strengths and weaknesses are. Now is the time to develop short-term action plans to address the most pressing matters and to fill critical gaps in the company's environmental capacity. Setting out a clear and tight timetable of no longer than six months for this phase will help to galvanize action across the company.

3. PILOT PROJECTS

In 2004, Citigroup and Environmental Defense joined forces for a five-week test of some new paper policies. In a handful of offices, the company bought 30 percent recycled paper for printers and made double-sided copies the standard. The test alone saved ten tons of paper, $100,000, and twenty-eight tons of greenhouse gases from reduced energy in paper making. Numbers like these make environmental initiatives much easier to pitch to the whole organization.

Pilot projects also provide a good way for a service business to enter the environmental fray. For many of them, pollution control and natural resource management issues can seem remote. But a well-designed "get your feet wet" exercise can help to highlight the company's connections to environmental challenges—and get people into the spirit of Eco-Advantage.

We've seen companies executing these pilot projects at all scales.

Small initiatives like Citigroup's recycled paper project can rack up early successes. Medium-term investments are useful for vetting ideas that require big financial investments. IKEA and Wal-Mart, for example, have both created model eco-stores that try out energy-saving techniques. Other pilot projects can take years. Unilever runs test farms all over the world with key crops such as peas, palm oil, and tomatoes. The company is experimenting with low-impact farming methods, with very encouraging results. Some farms have doubled their yield while cutting pesticide use by 90 percent and water use by 70 percent. Armed with on-the-ground experience, Unilever can gauge the investment needed and expected returns—laying the foundation for a broader rollout.

MEDIUM-TERM ACTIONS: EMBEDDING AN ECO-ADVANTAGE MINDSET

While the short-term focus is on understanding environmental risks and opportunities, the medium-term agenda lets you leverage what you've learned and execute on it. In this phase, companies embed environmental thinking across the business. They advance their Eco-Tracking Tools and build an Eco-Advantage Culture. Remember the proverbial flywheel in Jim Collins's *Good to Great*? Here's where companies start leaning in, pushing hard, and gaining momentum. They develop a strategic approach to environmental issues, which makes spotting opportunities for Eco-Advantage a more natural part of the business.

During this critical middle period, implementation should center on five areas: eco-tracking and management systems, employee ownership, communications to and from the outside world, internal education, and over-the-horizon scanning.

1. ECO-TRACKING AND MANAGEMENT SYSTEMS

Once you make the environment a business focus, eco-tracking and measuring progress become essential, as does development of a good environmental management system. Start by ensuring that you track compliance with all relevant laws. Recall GE's PowerSuite, which keeps an eye on a full range of issues company-wide, or by region, country, division, facility, and even production line. Be sure you're tracking regulatory requirements. Then move beyond compliance to track outcome metrics, such as greenhouse gas emissions, air pollutants, water use, and waste. Finally, add indicators that fit the company's culture to provide benchmarks and drive performance. DuPont's "strategic value added per pound of product" helps the company focus on how to create more value with less stuff.

The medium-term is also a good time to start collecting available information on supplier performance and building a materials database for key products. As most companies know, effective large-scale database projects can take years, not months, and usually cost more than what's budgeted. So start sooner rather than later.

Data-driven environmental management is becoming the norm. Companies that use indicators and information technology strategically stay a step ahead of the competition. They know where the real life-cycle impacts are, where their weaknesses and strengths lie, and where to find opportunities to help customers. Good data lays the foundation for generating Eco-Advantage.

2. ENGAGEMENT AND OWNERSHIP

When we gauge a company's progress toward an Eco-Advantage Mindset, one of the first things we look for is accountability. Do executives up and down the line feel ownership of environmental goals? Does their environmental performance affect their compensation?

The bonus and performance review system should reflect environmental thinking in some way. As we've noted earlier, many Wave-Riders tie compensation to environmental key performance indicators. Up to 25 percent of a manager's bonus can ride on these KPIs. The exact amount isn't vital, only that it's a noticeable part of the equation. A few leading companies find their cultural commitment to environmental issues so strong that they don't need to make it a part of the monetary reward system. But even in those cases, job reviews and career prospects reflect commitment to the company's values—you get promoted only if you "get it."

WaveRiders structure incentives to advance environmental strategy. They build environmental elements into job descriptions, regular performance reviews, bonuses, and awards.

3. EXTERNAL COMMUNICATIONS

Regular communications with outside advisors and critics can keep a company abreast of environmental developments. Experts will highlight some of the issues that are on the rise or should get more attention in the AUDIO analysis. Reaching out to people beyond your comfort zone will give you a first-person heads-up on complaints before someone launches a nasty protest or hostile web campaign.

NGO contacts and independent perspectives do more than control downside risk. They can help companies benchmark their performance, find Eco-Advantage opportunities, and develop creative solutions to problems. Consider setting up an Environmental or Sustainability (or Stewardship or Corporate Social Responsibility) Advisory Board. When Coca-Cola ran into a series of high-profile environmental problems over water use in India and refrigerants at the Olympic Games, one of us (Dan) helped then-CEO Doug Daft to create such a board. The group included energy maven Amory Lovins, green design expert Bill McDonough, and other experts from nearly every continent. The members meet twice annually to review Coca-Cola's environmental performance and help the company scan the horizon for emerging issues and concerns.

Learning from outsiders helps to fill the in-box with ideas, but companies need out-going messages, too. An environmental or CSR report is one step. For big enterprises, such reports are essential, but even for small and mid-sized companies, the exercise of producing one can help to focus attention, highlight issues, and bring opportunities for Eco-Advantage to the fore. The report should cover key metrics over time, both in absolute and relative terms. Paper copies are unnecessary if a company does a good online version, with the added advantage that an e-report can be continuously updated.

Finally, regular communications with some specific stakeholder groups is vital. Dell now has quarterly meetings with key players in the Socially Responsible Investment community. Chairman Michael Dell or CEO Kevin Rollins personally attend at least annually. These meetings mirror the long-standing practice in publicly held companies of holding quarterly analysis calls. Likewise, Northeast Utilities hosts an annual NGO day when they invite prominent environmental and community groups in for lunch and conversation.

4. INTERNAL COMMUNICATION AND EDUCATION

Of the twenty stakeholders in our players wheel, the most significant might well be employees. They make or break not only environmental initiatives, but the company's future in general. To succeed they need inspiration, information, and the right set of tools. Most WaveRiders do a good job of "knowledge management." Whether it's an intranet with examples of eco-efficiency from facilities around the world or more formal multiday training sessions, they spread the word on best practices. Companies like 3M and Northeast Utilities also help middle management deal with the pressures they face with special training on "hard choices."

Niche communications to a narrow internal audience also can be extremely useful. Dell, after taking it on the chin for a few years, has ramped up its stakeholder engagement efforts from virtually nothing to world-class in short order. Executives realized that part of the company's challenge stemmed from a general lack of knowledge within the company about these outside audiences. The solution was a quarterly newsletter sent to internal departments, as well as outside interests, describing the objectives, activities, and tactics of different stakeholder groups. Spreading this knowledge widely, Dell believes, will help its employees make better business decisions that keep stakeholder needs in mind.

Don't forget to encourage employees to read your environmental and CSR reports. Workers enjoy seeing what the company is doing right and hearing an honest appraisal of what's not going well. Giving employees something else to feel good about drives passion and morale in the workplace.

5. OVER-THE-HORIZON SCANNING

Shell's Scenarios group is perhaps the most famous example of corporate long-term thinking. These people think big thoughts, ask about alternative futures, and map out what the world might look like in the hydrogen economy or if geo-politics go sour. Very few companies need a group of this scale and breadth of perspective. But most companies could benefit from some formal process for looking over the horizon.

At the very least, a company should periodically pull together people from all over the organization and think about the big picture. In particular, managers working abroad often know about regulations and other trends that may change the global marketplace. Marketing executives may spot evolving consumer preferences. And purchasing managers may have a perspective on what's going on with suppliers. All this vital intelligence will be for naught unless somebody is given the responsibility for pulling it together.

The scanning exercise should address any issues highlighted in the AUDIO analysis but also consider longer-term pressures and market dynamics. AUDIO focuses specifically on direct impacts on the company and its value chain, while the over-the-horizon analysis looks for business drivers, trends, and developments that could shift entire industries or remake markets.

What you're shooting for is advance warning about new issues and a rough assessment of the impact these variables could have on the company's markets, financial position, and assets. What might climate change do to our competitive position? Are we contributors to the problem or possible solution providers?

In 1996, mining giant Rio Tinto held the first of many planning sessions on business drivers, which, as Dave Richards says, "could hurt us if ignored and be our friends if we manage them well." The team identified a series of growing problems that could constrain company growth—including water, human rights, and biodiversity—and launched Rio Tinto's efforts to mine with a lower impact on the land.

Scanning can take a company to some potentially surprising places. Remember McDonald's and the mercury button batteries. The first meeting where someone mentioned those as a potential problem for a company famous for Big Macs must have seemed a bit odd. But the long-range thinking helped McDonald's avoid a public relations nightmare. Or take Dell, where top managers were blindsided by the intensity of an aggressive NGO campaigning against e-waste. Never again, they vowed. So they've started exploring seemingly secondary issues, such as the environmental effects of the mining that produces the precious metals in their products.

In over-the-horizon scanning, look for items on the public agenda that seem to be increasing in intensity, even if they are only tangential to your business. Tracking second- and third-order connections can help you spot emerging business drivers. Brainstorming about how to respond to these trends can stoke innovative thinking about cutting costs, reducing risks, launching new products or services, and increasing intangible value.

LONG-TERM ACTIONS: MAKING ENVIRONMENT A CORE ELEMENT OF CORPORATE STRATEGY

As an Eco-Advantage Mindset takes hold, a few more steps emerge that can carry a company's thinking to a new level: (1) supply chain auditing, (2) rethinking products and reexamining markets, and (3) building partnerships with key external stakeholders. These three steps drive environmental thinking deep into business strategy and sustain Eco-Advantage for the long haul.

1. SUPPLY CHAIN AUDITING

A lot of big companies review their supply chains for environmental and social issues—or so they claim. For many, the "audit" is little more than a formality that creates paperwork for suppliers. The auditor asks a few basic questions, often narrowly focused on whether the suppliers are complying with the law, and that's where the review ends. Yes, pressure back through the chain to make sure everyone is following the law is a good thing. But focusing on compliance isn't enough today—and certainly won't generate Eco-Advantage.

As we described in Chapter 7, IKEA has one of the most thorough supplier audit programs in the world. The company dedicates dozens of employees and millions of dollars to diving into the supply chain and asking detailed questions about environmental and social performance. For IKEA, the IWAY program not only serves to flag brand-threatening risks, it also helps managers understand their business better.

Auditing the supply chain is a challenge. Unless you have real market power, it's hard to get all the information you'd like. But as efforts on this front heat up, smaller companies have an opportunity to piggy-back on what the big brands are already doing. If a company buys from an IWAY-compliant supplier, for example, it can be sure that company has been checked out thoroughly.

2. RETHINKING PRODUCTS AND REEXAMINING MARKETS

In the long run, making your own operations as eco-efficient as possible could be the margin of survival in a cost-conscious world. Concentrating on eco-efficiency almost inevitably raises fundamental questions about markets, products, and services. Can we storm the marketplace by taking advantage of changing environmental circumstances or regulations? Can we create breakthrough products to cut our customer's environmental burden or capture new green customers? Or entirely reconceive the way we operate to improve resource productivity and slash costs? The answers to those questions are what Eco-Advantage is built on.

But let's face it—sales growth is what gets CEOs and shareholders excited. New products with an eco-twist can grab customer attention and move the top line. WaveRiders have learned that you can turn green into gold when you cross-breed environmental awareness with good design.

In this spirit, Intel co-located designers and environmental professionals as an innovation strategy. At a more mundane level, IKEA developed a simple graphic of four life-cycle phases—raw materials, manufacturing, use, and end-of-life—which it calls the eWheel. IKEA product designers use a list of twenty-five key questions focused on these four stages, such as: Did I minimize use of chemicals like glue and paint? Can we make this with suppliers who are rated the highest in the IWAY supplier program? Is the product easy to disassemble? Similarly, Herman Miller guides its designers toward eco-friendly results with its simple "red, yellow, green" system. Based on the materials they choose, each product gets a total score—more points for green materials—which gives designers something to shoot for.

Going past redesign to reimagination can be even more exciting. 3M's famous "work on what you like for 15 percent of your time" rule is the kind of organizational guideline that helps. Google has picked up on this tactic, and has its own 10 percent rule for following "unrelated projects" and wild, out-there ideas. It's clearly working for them. Giving employees these open spaces to rethink how the company, industry, or world works is a great way to push through boundaries and inspire employees. It takes time to build a culture that spurs innovation, but it's almost always worth it.

3. STAKEHOLDER MANAGEMENT AND PARTNERSHIPS

More than one WaveRider told us that they started out with a dismissive or defensive attitude toward NGOs and other outside groups. Once they began to meet with stakeholders, though, they realized the value of reaching out and seeing their companies through the eyes of others. Partnerships with NGOs, communities, and other organizations broaden understanding and provide a useful mechanism for feedback and learning.

Despite the diversity of players in a Green Wave world, governments remain a powerful force defining the marketplace and the rules of the game. But the roles these officials play are evolving, and smart companies are watching for new opportunities to engage.

Connections to idea generators—academics, think tanks, and research centers—can also pay big dividends in a knowledge economy. WaveRiders are launching joint ventures of all types with these players, using these relationships as a source of new ideas, gaining insights into the direction technology is moving, and benchmarking their efforts against cutting-edge thinking.

If problems are large scale, long term, or require a major investment in infrastructure, industry partnerships might make the most sense. Sometimes it's worthwhile to pool your clout and share knowledge. For example, German companies found it useful to work together to develop the Green Point system, a private waste collection infrastructure they needed to meet the country's stringent take-back laws for product packaging. When the burden of handling an issue company-by-company will weigh down a whole industry, working together may be the right answer. There can be strength in numbers.

Steps to Develop a More Complete Stakeholder Strategy

As challenging, and interesting, as our earlier mapping exercise may seem, it's really only the first step in a comprehensive stakeholder engagement strategy. Our two-by-two matrix is a way to brainstorm about who the relevant groups and individuals might be. It forces a systematic review of the 20 categories to help identify gaps.

A complete stakeholder approach goes much further. It starts by helping to alert you to the full range of groups and actors that could impinge on your business. It then provides a way to hone in on key entities to track more carefully and perhaps engage with more deeply. And it offers a mechanism for identifying and evaluating potential partnerships. The following steps form a funnel that allows thinking to start broad and then narrow:

1. **Brainstorming.** As we suggested as a short-term action, it's critical to figure out which groups work on the issues you face. You want to identify friendly stakeholders—and those who are less so. As silly as it sounds, Googling your company's name can locate many pockets of concern. Has someone launched a website with a URL that reads something like [yourcompany]sucks.com? If so, who did it and what's the beef? What are the major NGOs saying about you or your industry?

2. Grading and scoring key stakeholders. To determine which entities or groups deserve focus, we suggest rating stakeholders, particularly NGOs, on three dimensions:

- Power and influence
- Credibility and legitimacy
- Issue urgency both to you and the stakeholder

3. Stakeholder engagement evaluation. The most important stakeholders—those whom you grade highly on two or three of the di-

Stakeholder Engagement Evaluation Matrix

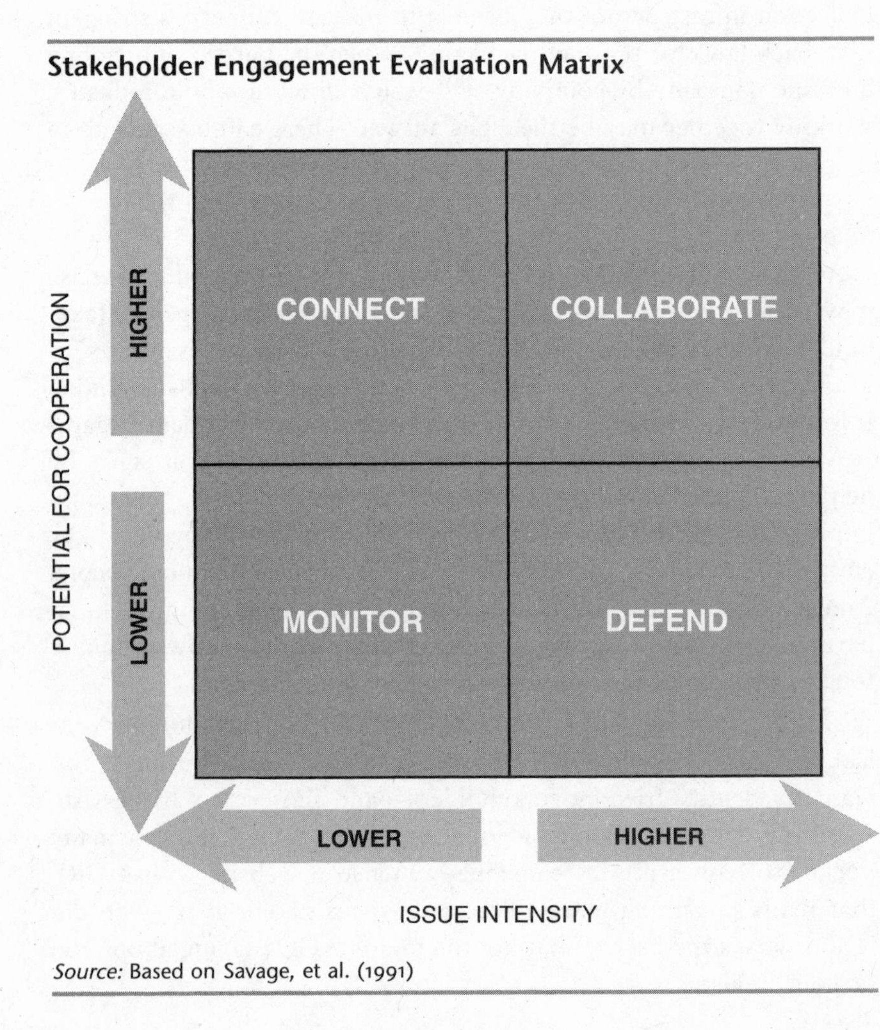

Source: Based on Savage, et al. (1991)

mensions—require further attention. To determine how best to handle these entities, another simple two-by-two matrix may prove to be useful (see figure). It provides a way to map stakeholders based on: (a) how cooperative (as opposed to confrontational) the group is and (b) the intensity of your interest in the issues they address. This matrix clarifies what sort of engagement is optimal for each group. Should you just monitor their activities? Try to connect with them? Work to defend against them? Or take the leap and collaborate with them?

4. Compatibility assessment. The short list of organizations and people in that top right box provides a good starting point for thinking about possible entities with which to work closely or even partner on a project. Before contemplating a full-fledged partnership, however, it's useful to evaluate rigorously how compatible the group would really be. What is their over-arching vision? On what issues do they focus? What is their general mode of operation (conflict or collaboration)? Does their style and organizational culture match yours? Do their strengths complement yours?

5. Due diligence on potential partners. Once you've identified a compatible partner, you need to dig deeper. Is their leadership strong? Are their finances stable? Who funds them? Are they transparent about their funding and governance? How effective are they at achieving their goals? Do they seem like good people to work with?

6. Determination of partnership strategy. Before launching a partnership, you'll need to develop a game plan for success. What are the goals of the joint venture? Who will do what in the partnership? Are responsibilities clear? Have the necessary resource commitments been spelled out? What does success look like?

This may all sound a bit dizzying. There is, however, no real choice. Simply put, we live in a world of rising transparency and emboldened stakeholders. Corporate success now requires careful attention to a wide range of relationships.

THE ECO-ADVANTAGE BOTTOM LINE

We suggest breaking up the action plan to promote Eco-Advantage into the short-term (0 to 6 months), medium-term (6 to 18 months), and long-term (18 months and beyond). Short-term actions should focus on baseline analyses and first steps toward developing an environmental strategy:

- AUDIO analysis—Issue spotting
- Stakeholder mapping
- Core capabilities assessment
- CEO statement of commitment
- Priority action plans
- Pilot projects

Medium-term actions should seek to embed an Eco-Advantage Mindset in the company culture:

- Setting up eco-tracking and environmental management systems
- Promoting engagement and ownership
- Developing external communications and outreach
- Strengthening internal communication and education efforts
- Over-the-horizon scanning

Long-term actions should center on making environment a core element of strategy. This will require advanced Eco-Advantage tools and exercises:

- Supply chain auditing
- Rethinking products and reexamining markets
- Stakeholder management and partnerships

Chapter 12 Eco-Advantage Strategy

In Michael Porter's highly regarded strategy model, companies gain competitive advantage by lowering costs or differentiating products. But today the traditional points of competitive differentiation are being squeezed on all sides. Outsourcing—and the lower labor costs it promises—is available to almost any business, big or small. Other once unassailable sources of advantage, such as access to capital or low-cost raw materials, are disappearing as markets go global. Competitive advantages are becoming ever more difficult to establish and maintain.

This restructured landscape requires refined business strategy. The capacity for innovation—bringing imagination to bear to solve problems and respond to human needs—lies at the heart of success. Companies must find new ways to break out of the pack. Those that don't will struggle to keep up in the marketplace.

Environmental strategy offers just this sort of opportunity. As a relatively new variable in the competitive mix and a market-reshaping issue, the environment presents a lens through which to examine a facility, company, or industry—and a way to bring fresh thinking to bear. Careful use of the environmental perspective can help to reduce costs and risks. But it also can drive upside gains, increasing revenues and the value of hard-to-measure but important intangibles such as reputation. Finding new market spaces, satisfying customers' needs in new ways, and just plain doing the right thing—which many important stakeholders appreciate and reward—all have the potential to add real value.

The business world is waking up to an inevitable and unavoidable truth: The economy and the environment are deeply intertwined. All goods depend on the bounty of Nature and the services it provides. Without careful stewardship, natural resource constraints will encroach on a growing number of companies and industries. Concern about these trends is driving laws, rules, and expectations that will further restrain business. The environment thus ranks as a macro-issue right up there with globalization, the Internet, and the other megaforces that keep CEOs up at night. In this new, more complicated and interconnected world, environmental strategy emerges as a critical point of competitive differentiation.

In the very near future, no company will be positioned for industry leadership and sustained profitability without factoring environmental issues into its strategy.

Strategy no longer rests in the hands of narrowly focused planning teams. Today, every company's financial future depends on executives who possess the ability for integrated thinking. The companies "in the barrel" of the Green Wave adeptly incorporate the environment into their core strategy. They work with a dynamic and holistic vision of how a company operates and engage the full range of stakeholders who can shape the company's future. They create enduring

Eco-Advantage by thinking differently, adopting tools to understand their companies' environmental challenges and opportunities, and embedding attention to stewardship in their corporate values.

We see four foundational elements—an Eco-Advantage Mindset, Eco-Tracking, Redesign, and Culture—underpinning environment-driven innovation. In this chapter, we review how to develop these critical supports. We also explore the forces influencing companies, the Green-to-Gold Plays that offer a way to get ahead of the competitive pack, and the hurdles to avoid on the way to Eco-Advantage.

We bring these elements together in a complete picture of environmental strategy (see figure). The adept use of the Eco-Advantage Toolkit drives successful execution of the Green-to-Gold Plays. Natural forces and a range of players exert significant influence on the process. Companies face many hurdles and risks of failure on the

Eco-Advantage Strategy

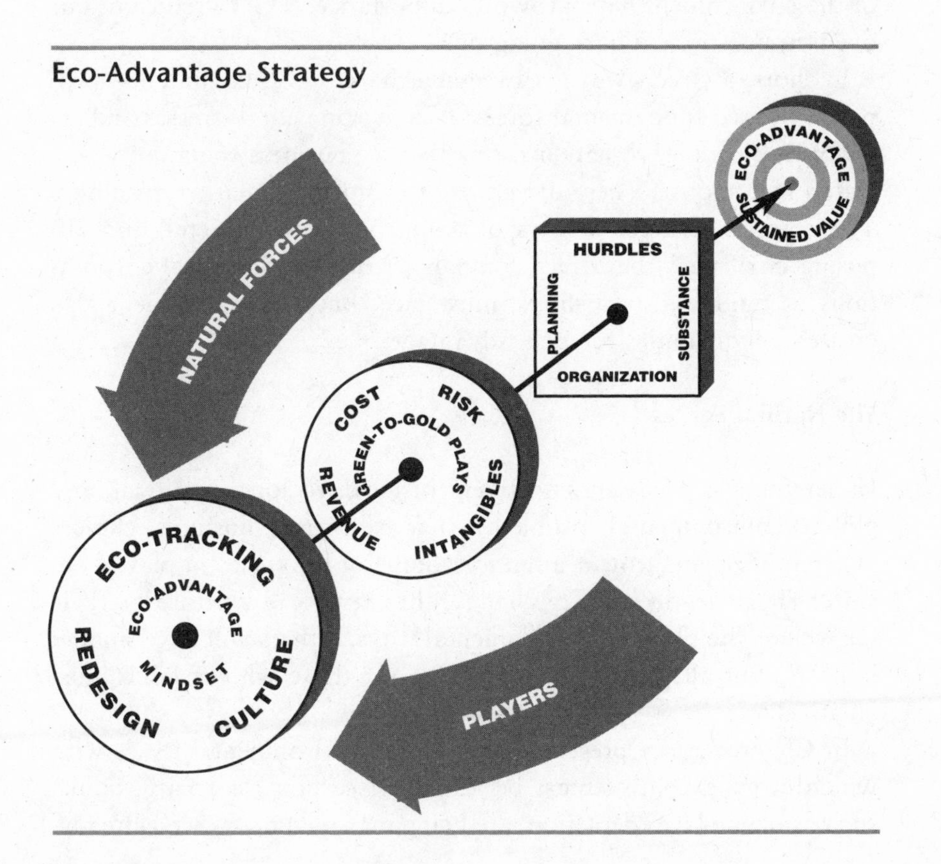

way to Eco-Advantage. But those who persist and learn from experience find ways to innovate, create value, and build competitive advantage.

THE PRESSURES: NATURAL FORCES AND PLAYERS

We started this book with stories of companies at various stages of dealing with environmental issues. Sony, having learned a hard lesson with its PlayStation game systems, spent over $100 million building a supplier audit system to catch problems before they emerge. BP, while it was "looking for carbon," found a range of efficiency gains and discovered an astonishing $1.5 billion in value waiting to be unlocked. And Wal-Mart, GE, and Goldman Sachs, in moves that the companies' previous leaders might find surreal, have launched major environmental initiatives. After years of neglect, considerations of the environment have grown in importance. So let's return to our opening question: What's going on?

In short, a Green Wave is sweeping the business community, propelled by two fundamental forces: (1) environmental stresses and (2) a world of people who are insisting that the business community take action in response. These drivers are transforming market dynamics. They are rendering old ways of doing business obsolete—and imposing challenges that every company, from multinational corporations to mom-and-pop shops, must face. But this realignment also creates opportunities for Eco-Advantage.

The Natural Forces

Under the Wave lies an assortment of local, regional, national, and global environmental problems that constrain business choices and require management attention. Some of the issues in play, from water shortages to climate change, threaten to restructure markets (as well as the planet) in fundamental ways. Others will have smaller impacts. But all provide opportunities for those who respond most creatively.

In Chapter 2, we presented ten critical environmental issues with which every executive must be familiar (see box for recap). Some, such as ozone layer depletion, are being managed pretty well already.

TOP 10 ENVIRONMENTAL PROBLEMS

- **Climate Change**: The build-up of greenhouse gases in the atmosphere threatens to lead to global warming and the accompanying rising sea levels, changed rainfall patterns, and increased intensity of storms.
- **Energy**: A carbon-constrained future will require a shift toward new modes of power generation and sustainable energy or new technologies for cleanly burning fossil fuels.
- **Water:** Water-quality issues and water shortages are threatening licenses to operate and constraining business activities all around the globe.
- **Biodiversity and Land Use**: Ecosystems play a critical role as life support for both humans and nature. Unmanaged development undermines this capacity through habitat destruction, loss of open space, and species decline.
- **Chemicals, Toxins, and Heavy Metals**: These contaminants create a risk of cancer, reproductive harm, and other health issues in humans, plants, and animals.
- **Air Pollution**: Smog, particulates, and volatile organic compounds pose a risk to public health, especially in the developing world where trends are worsening. Indoor air pollution is now recognized as an added problem.
- **Waste Management**: Many communities still struggle with the disposal of their solid and toxic waste, especially in countries that are industrializing and becoming more urban.
- **Ozone Layer Depletion**: Depletion has been substantially reduced by phasing out CFCs, but some substitutes continue to cause thinning of the Earth's protective ozone shield.
- **Oceans and Fisheries**: Overfishing, pollution, and climate change have depleted fish stocks and damaged marine ecosystems across the Seven Seas.
- **Deforestation**: Unsustainable timber harvesting plagues many parts of the world, leading to soil erosion, water pollution, increased risk of flooding, and scarred landscapes.

Others verge on crisis. The most pressing issues, like climate change, promise to affect every business, large and small, in every industry. Others are strategically important for business only in certain contexts. And there are a dozen more issues with industry-specific impacts that could also be significant.

What makes these pressures a matter of business strategy is one simple truth: Our economy rests on the asset base of the natural world, not the other way around. Where resources are threatened, ripples will move through society and across the corporate world.

These issues are complicated. The fact that the underlying science is often complex, even contradictory at times, makes the situation even more volatile and environmental policymaking highly contested. Some public policy choices could drive whole industries into obsolescence. Can the coal business survive in a carbon-constrained world? We can be sure that those affected will not go quietly. In fact, coal companies are struggling mightily to resist pressures for action on climate change.

On the other side of the ledger are companies and industries that will be crippled if we *don't* take action to stem environmental losses. The skiing industry won't suffer alone if global warming hits as hard as many fear. And everyone's insurance costs will rise if reinsurers jack up rates to handle the growing claims from climate change–related natural disasters.

The diversity of issues, variety of interests, and range of scientific uncertainties can be daunting. But executives can't just throw up their hands in despair and confusion. Companies need to stay on top of these challenges no matter how complex, both to reduce risk and to pounce on emerging opportunities.

SCIENTIFIC UNCERTAINTY

After a CEO pow-wow on climate change, Cinergy CEO Jim Rogers said, "Forget the science debate. The regulations will change someday. And if we're not ready, we're in trouble." He's voicing a tough reality. Even without absolute scientific certainty, the list of issues that business can't avoid is growing. The reality is that environmental problems will find you whether or not you care to be inconvenienced by them.

The Players

An evolving set of stakeholders who care about these issues and wield influence over corporate behavior are adding to the momentum behind the Green Wave. In the traditional B-school strategy, companies are led by charismatic leaders who make bold decisions, influenced by only a handful of actors: competitors, customers, "channels" (suppliers and distribution), and maybe government regulators. And all of these players are subservient to the almighty shareholder.

The balance of power hasn't entirely changed, but CEOs have cause to feel like the ground is shifting beneath their feet. New stakeholders are asking tough questions about social and environmental performance. Civil society in general, and environmental NGOs in particular, have emerged as forces to be reckoned with. Coordinated action against irresponsible corporations has never been easier thanks to e-mail, the Internet, and other modern communications technologies. And activist shareholders, including large mainstream investment companies, have suddenly found their voice. Nervous boards now watch over the shoulders of CEOs like never before.

As difficult as it is, ignoring these drivers of today's business reality is even more unwise. Sure, a few remaining corporate titans still think that most environmental issues are over-blown creations of some Birkenstock-wearing, tree-hugging fringe element. As misguided as this caricature may be, the truth is actually beside the point. The concern around these issues is broad enough that every director, executive, and manager must pay attention to the Green Wave. Anyone who thinks he or she can avoid it risks being drowned by the growing array of legal requirements—like those laid out in the Sarbanes-Oxley law—that mandate attention to potential "material" issues, including environmental challenges.

Our research has identified an array of twenty Green Wave players, some marginal for the time being, but many more growing in power and importance. These stakeholders break into five groups:

Rulemakers and Watchdogs: NGOs, regulators, politicians, and the plaintiff's bar. The array of players in this category is expanding both "vertically" and "horizontally." National regulators in many regions, particularly the European Union, are growing more aggressive. But down the vertical scale, regional, state, and local governments are getting involved in environmental issues in ways they never have before. American states, for instance, are setting their own renewable energy goals. And U.S. mayors are agreeing, on their own, to implement the climate change emissions reductions targets of the Kyoto Protocol. Go up the vertical dimension and you'll see continent-wide regulations like the EU's laws on chemicals and electronics. At the largest scale, calls for global regulations on a range of environmental issues are growing.

On the horizontal dimension we see an increasing breadth of power centers. Companies must track not only the requirements of governments but also the demands of an incredible diversity of NGOs and other self-appointed watchdogs (like bloggers). These new players can do great damage to a company's reputation quickly. And the Internet only makes it easier to turn the slightest corporate misstep into an international incident. But the news is not all bad for companies. An explosion of partnerships has created dynamic alliances between former adversaries. NGOs are working *with*, not *against*, business like never before.

Idea Generators and Opinion Leaders: Media, think tanks, and academics. In today's innovation-driven world, being connected to those creating knowledge, launching ideas, and shaping the political dialogue provides a competitive leg up. To tap top thinkers, WaveRiders have launched partnerships with academic institutions and research centers around the world.

Business Partners and Competitors: Industry associations, competitors, business-to-business buyers, and suppliers. Businesses are now finding ways to work together for environmental gain and to get ahead of the issues. The trend started twenty years ago with the chemical industry and its Responsible Care program and has blossomed since.

Electronics companies have joined together to set supply chain standards. Big energy users formed the Green Power Market Development Group to encourage renewable energy development. A range of paper buyers, from Staples to Toyota, formed the Paper Working Group to coordinate the requirements they'll ask paper suppliers to meet, in effect imposing a privately determined set of environmental standards.

The pressure for change usually starts with big brands, but the effects are felt on obscure midsized companies caught in the web. A few years ago, Home Depot changed its procurement policies to eliminate products that originated in endangered forests. In response, the flooring company Romanoff changed one of its products from plywood to renewable wheat straw. As President Douglas Romanoff said, "The Home Depot purchasing policy . . . has produced a direct ripple effect that has resulted in a significant change in the material we will use in the future."

Consumers and Community: CEO and executive peers, consumers, kids, and communities. When an environmental boycott works, it's impressive. In the wake of the famous Brent Spar oil platform incident in the mid-1990s, a million Europeans cut up Shell credit cards. The company took notice and set off on a decade-long quest to improve stakeholder relations. And the list of those who might call you out is expanding.

Investors and Risk Assessors: Employees, shareholders, insurers, capital markets, and banks. As every executive knows, in today's knowledge-based economy, capturing the best and the brightest is not just helpful, it's essential. Before the most talented workers invest their time and energy in a company, they increasingly ask what a potential employer stands for. They want to work for companies whose corporate values are in harmony with their own worldviews.

Traditional investors have emerged as the new 800-pound gorilla in the environmental space. Led first by insurers and now joined by major banks, the financiers have started to look hard at environmental risks and liabilities. Over forty banks have signed on to the Equator Principles, which demand thorough environmental reviews before loans are approved. But the Principles are just the starting point. Goldman Sachs, JPMorgan, Citibank, and many others are wrapping environmental considerations into lending decisions in dramatic new ways.

ABN AMRO, one of the founders of the Equator Principles, has developed a new way of looking at its portfolio of loans. The company charts borrowers on a classic two-by-two matrix, with capacity to handle and mitigate environmental risks on one axis and commitment to do so on the other. In the near future, ABN AMRO hopes to graph all potential loans against these criteria. Upper-right quadrant loans—with borrowers who "get it" and have the means to fix any environmental issues—would be no-brainers. Deals falling in the upper-left or lower-right quadrants would require some work on the borrower's part. Loans to those in the lower-left corner—the unwilling and unable—would not be made.

ECO-ADVANTAGE TOOLKIT

To be successful in integrating the environment into business strategy, companies need to cultivate an Eco-Advantage Mindset backed by a set of tools, including Eco-Tracking, Culture, and Redesign.

Eco-Advantage Mindset

During our interviews with dozens of companies, we looked for what made them effective at bringing environmental considerations into the mix with other business goals. We found an overarching set of five principles that guided their thinking.

1. **Look at the forest, not the trees.** WaveRiders think broadly about time, payoffs, and boundaries. They make decisions with the long-term in mind, positioning themselves for a tighter regulatory framework, rising consumer expectations, and market realignment driven by natural constraints. WaveRiders also include intangible benefits in their payoff calculations. They put a value on things like lower risk, higher employee retention, stronger customer loyalty, and bolstered brand value. Finally, they think beyond their own operations and look at the whole value chain, from raw materials and suppliers to customer environmental needs and desires to product end-of-life.
2. **Start at the top.** All the WaveRiders have "C-level" support for their efforts to seek Eco-Advantage. Start with CEO commitment. That alone won't put you over the top, but no company will get far without it.
3. **Adopt the Apollo 13 Principle.** Don't take no for an answer. Companies and industries have time and again shown incredible creativity in solving seemingly intractable environmental problems. WaveRiders focus on innovation and getting people to use an environmental lens to think about their work in a new way.

4. **Recognize that feelings are facts.** Leading companies know that they have to deal with what communities, NGOs, and other stakeholders *feel* the big environmental issues are. Instead of blindly defending their own position or downplaying others' concerns, they recognize the need to meet people where *they* stand. They don't let outsiders dictate the agenda, but they do establish a dialogue with friends and foes alike.
5. **Do the right thing.** It's amazing how often we were told that a sincere belief in doing what's right was behind WaveRider decisions. Values do matter.

Eco-Tracking

The next element of the Toolkit is a systematic approach to capturing and using good information. WaveRiders use issue-spotting tools like the AUDIO analysis as well as Life Cycle Assessments to understand their environmental impacts. They look at the eco-consequences of their products all along the value chain, upstream and downstream.

These tools are most effective when they rest on a foundation of good data, careful planning, and an environmental management system. The best systems track dozens of metrics by region, division, factory, even down to particular production lines. And they track the same metrics globally. Alcan keeps a worldwide data warehouse on environmental performance. That way, all divisions work in a consistent way and are assessed against similar measures. Common data helps headquarters benchmark performance, set targets, and monitor progress closely. As we mentioned in Chapter 7, we recommend a core set of metrics that track results on energy use, water and air pollution, waste generation, and compliance.

Companies also need outside perspectives to learn where they stand in the world. Many WaveRiders establish relationships with environmental experts. Some create advisory boards to "peer review" their environmental efforts and stay ahead of issues that could slam them. Some invite the fox into the hen house by partnering with NGOs and other critics. No one necessarily looks forward to spending time around a table with those who have raked them over the coals in the past, but their feedback is a valuable form of Eco-Tracking.

In the same spirit, WaveRiders also reach beyond environmental groups to work with communities, governments, other companies, and any other stakeholder than can provide them with credible information on environmental issues and changing market conditions.

***What gets measured gets managed. Knowledge is power.* Trite phrases perhaps, but the leading companies treat these ideas not as throw-away lines but as calls to action. They leverage data and knowledge to generate sustained marketplace advantage.**

Redesign

Tracking data helps define the playing field, but companies gain an Eco-Advantage only when they understand environmental market drivers, use their knowledge to drive innovation, and change products and processes. WaveRiders redesign their products, the spaces around them, and even their supply chains.

Eco-design, the second point of the Eco-Advantage Toolkit, has helped companies like Intel and Herman Miller design out environmental problems before they arise, saving time and money down the road. Factoring environmental considerations into product design also means helping customers reduce their environmental footprints. In a world of rising energy prices, for example, having the most energy-efficient product on the market will often translate into rising market share.

A number of leading companies have embraced green building as well. Why? Because well-designed, energy-efficient facilities often save money, improve worker productivity, and send a signal about corporate values.

A few companies such as IKEA are going beyond compliance audits and pressuring suppliers to change business practices. These best-in-class companies are redesigning entire value chains to reduce environmental and social impacts.

Culture

The third leg of the Eco-Advantage support structure centers on building a corporate culture that promotes environmental thinking and innovation. While every business is unique, we found four common approaches across WaveRiders:

1. **Stretch goals.** WaveRiders use targets that seem symbolic and even uncomfortable, but inspire innovation and a broad reexamination of how they do things. "The goal is zero" is one example. A few companies even discover that "zero" is not an unreachable number.
2. **Decision-making tools.** Top-tier companies refine traditional cost-benefit analyses to allow for intangibles. They tweak internal hurdle rates (or "pair" projects) to tip the balance in favor of some environmental investments. And they use Adam Smith's invisible hand, through internal markets, to guide decisions.
3. **Ownership and engagement.** CEO commitment gets the ball rolling. Engaging other senior managers and all employees keeps it going. WaveRiders use various tools to make executives sit up and take notice of environmental priorities. Some are soft, like assigning executives to be "stewards" of an environmental issue. Other tools have a decided bite to them. GE's Session E, most notably, asks plant managers to explain their environmental performance in front of bosses and peers.

   WaveRiders also drive interest and engagement by cross-fertilizing environmental and line managers. At IKEA, 3M, DuPont, and many others, the top environmental officials came out of line businesses. No division head can tell these experienced operational executives that they don't get it. The credibility of the environmental goals is greatly enhanced by sending the right messenger.

   Money and incentives focus attention, too. Many leading-edge companies build environmental key performance metrics into bonuses. At some of the globe's greenest companies—places like IKEA, 3M, and Herman Miller—deep cultural values, including a commitment to stewardship, motivate managers.

Finally, awards, even if only plaques, go a long way. Many WaveRiders have annual environmental or sustainability awards. Because the companies clearly treasure their green commitments, it's a real honor to be singled out for environmental successes.

Whether through direct pay incentives or cultural pressure, WaveRiders find ways to walk the talk and align their statements about environmental commitment with on-the-ground operational decisions.

4. **Storytelling.** Smart companies tell the stories of their environmental goals, successes, and lessons learned to nearly anyone who will listen. The knowledge-sharing can happen through an internal intranet of best practices or public reports. These documents inform all stakeholders, but particularly employees, about what the company is doing right—and wrong. For internal audiences, eco-training is an even more direct form of storytelling. Teaching line managers how to seek out eco-efficiency opportunities or taking middle managers through some what-if scenarios that demonstrate hard trade-offs can jump-start innovative thinking and better decisionmaking.

GREEN-TO-GOLD PLAYS

An Eco-Advantage Mindset, supported by the right tracking tools, a focus on redesign, and a culture of environmental stewardship, lays the foundation for turning green to gold. But the real action lies in the strategies that create value, the Green-to-Gold Plays.

Like any other business strategy, our Green-to-Gold Plays aim to reduce the downsides a business faces (cost and risk) or increase the upsides (revenue and intangible value). Unlike many others, though, these plays don't sacrifice responsibility in the pursuit of profit—or

Green-to-Gold Plays Framework

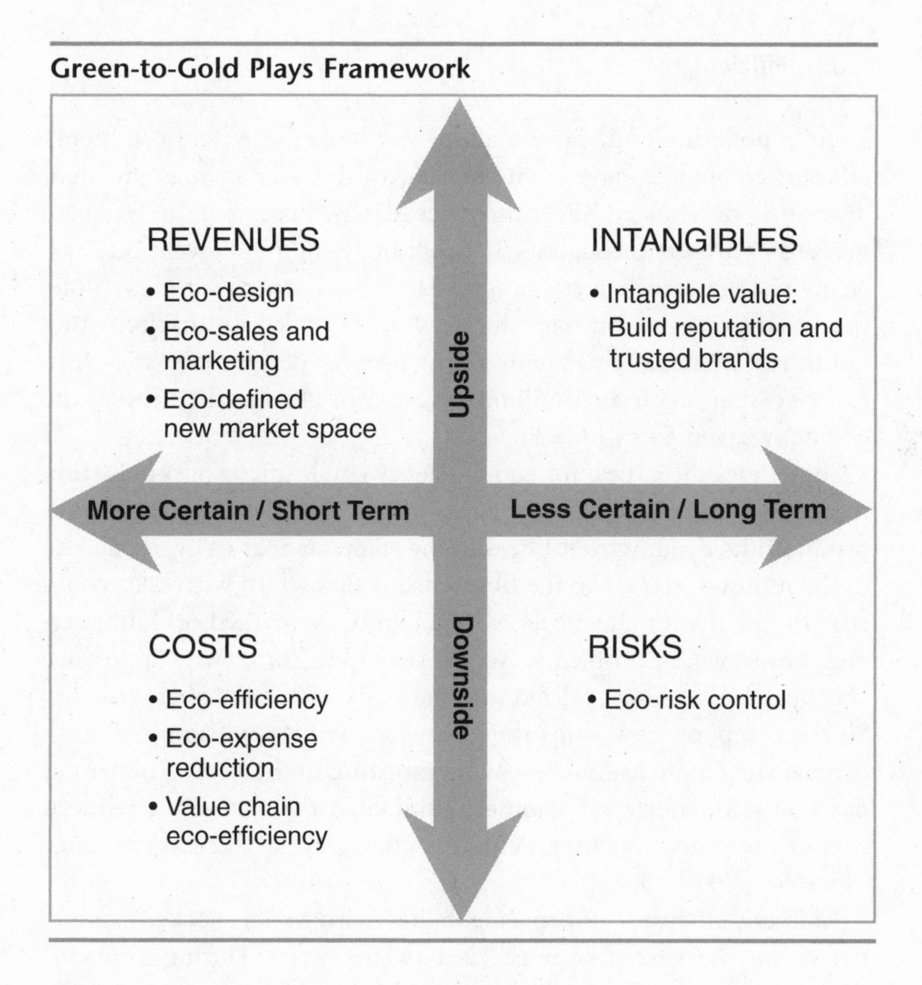

profit in the pursuit of responsibility. Our WaveRider companies offer proof every day that doing good and doing well can be symbiotic.

We've mapped the eight Green-to-Gold Plays drawn from our study of WaveRiders onto the two-by-two strategy framework we outlined earlier (see figure). Not surprisingly, most green business efforts to date have focused on the lower left box. Cost reduction is extremely low risk, easy to sell internally, and often pays back quickly. It can yield competitive advantage. But our research suggests that, by focusing solely on the cost side, many companies are missing chances to generate broader Eco-Advantage. Most companies have not yet executed all of the plays—they're leaving money on the table.

1. Eco-Efficiency

Cutting pollution and waste makes good business sense. Even highly efficient companies have been shocked to discover savings they had previously overlooked. Over three decades, 3M continues to find new ways to pare costs through its 3P program, Pollution Prevention Pays. Many changes can be very simple. STMicroelectronics, for example, put in larger air-conditioner ducts, which allowed its air-circulating fan to run more slowly. The fan now uses 85 percent less energy. In just one year, with $40 million invested in changes like these, the company saved $173 million.

Sometimes the search for eco-efficiency can leapfrog past reduction to outright elimination of a process or resource. Rohner Textil once produced its dyed, woven fibers in the same manner as everyone else in the industry. To make the fibers strong enough to weave, it would coat the yarn with chemicals, which had to be washed off later, creating wastewater problems. While searching for a way to reduce chemical use, Rohner realized that humidity makes the fiber stronger. So the company now skips the chemical coating and simply doesn't dry the yarn quite as much, leaving moisture in the fiber. Rohner cut out one step, shortened another, eliminated the chemicals, reduced energy use, and cut costs. A pretty good day at the eco-efficiency office.

Rohner's efficiency improvements have driven per worker productivity up 300 percent over the last twenty years. During a vicious downturn in its industry, Rohner, unlike many other companies, remained profitable.

Eco-efficiency simply depends on cutting out waste and using resources productively. Businesses that run lean are more productive, profitable, and less polluting.

2. Eco-Expense Reduction

Efforts to lower direct environmental costs such as landfill fees or regulatory paperwork can also return big dividends. DuPont has saved billions on pollution control, and that's only the *measurable* cost of waste. In one case, the company cut rejects from the Lycra production line from 25 percent of volume to less than 10 percent. That focus on reducing waste saved material, lowered landfill costs, and freed up $140 million in saleable product. It also meant the company could delay building another plant, saving many millions more in capital expense. The ripples from cutting waste and eco-expense can overflow and save money in many ways.

3. Value Chain Eco-Efficiency

Companies that look broadly for environmental gains and use tools like Life Cycle Assessment often find ways to reduce costs throughout their value chains. The play here is to try and capture that value, which can be a difficult task. In Chapter 4, we talked about one area in which companies are quite effective—distribution. IKEA and others stuff their trucks through smart package and product design, and save money.

4. Eco-Risk Control

With the rise of transparency, the risks to a business and its brand can come from anywhere. A substantial amount of goodwill is tied to corporate reputation. If a distant supplier dumps waste in a river or employs children, the major customer, with an international brand, may well be the one to pay the price.

WaveRiders identify potential risks and act on them as early as possible. When McDonald's pushes back on its supply chain to lower antibiotic use in chickens, or asks for documentation that ensures that cattle do not have mad cow disease, it's lowering the risk of contaminating its brand. Intel spends millions to ship its hazardous waste from some developing countries to the United States so it can be disposed of properly. Why? Intel doesn't trust the waste-handling system in some countries where it operates. And company officials know they'll be blamed if something goes wrong.

WaveRiders get ahead of regulations before they get tighter. BP began its Clean Cities program and sold cleaner-burning, lower-sulfur fuels in part to get out in front of more stringent air quality laws. "The driver was that sulfur regulations would come," BP's Chris Mottershead told us. "Rather than deliver on a regulated schedule, we decided to go early and try for a market benefit."

Anticipating regulations can put a company in a position to meet requirements at a lower cost than its competitors. Some companies have even obtained a competitive edge by lobbying for tighter controls. Remember, it's often the *relative* regulatory burden that matters.

5. Eco-Design

Redesigning processes and products to cut waste and pollution is a big part of Eco-Advantage. Keep in mind, too, that a great deal of potential gain might lie outside the factory gates. Helping customers reduce their environmental problems can strengthen customer loyalty and attract new sales. Reducing a product's energy use or toxicity also can add to customer value. Like Johnson Controls, which sells entire energy management systems, companies that find ways to lower customer burden can profit.

6. Eco-Sales and Marketing

Marketing the green qualities of products can drive sales. When Wausau Paper launched a new brand extension of "away from home" products—paper towels, toilet paper, and the like—it first certified the product line with Green Seal, an NGO specializing in environmental product labels. The company then rebranded the product EcoSoft Green Seal, putting the certification right in the name. In an industry growing only 2 to 3 percent per year, Wausau's sales in this market leapt 44 percent in the first two years.

In fact, Wausau took an unusual route by focusing its marketing pitch squarely on the environmental message. Products that scream "green" to the exclusion of other qualities often die on the shelves. As Shell learned with its Pura gasoline, a product often needs to stand on other attributes first before selling the environmental story. Green, we've found, is often best used as the "third button."

7. Eco-Defined New Market Space

Environmental vision can create new market space and value innovation. Toyota set out to redefine the 21st-century car and has come pretty close. Many customers now seek a hybrid, not a midsized car, and they'll pay a substantial premium or wait months for a Prius in particular. For these consumers, there is no substitute.

Successful, long-lasting companies regularly redefine themselves. Environment-inspired innovation offers companies a new and exciting way to find fresh expressions for their capabilities.

Looking for environmentally defined market space can seem to lead companies far afield. Take John Deere's recent foray into renewable energy. The tractor maker started up a business unit to help farmers harvest wind energy. Deere will offer financial backing and consulting. This may seem an odd fit, but we see it as an interesting play. A company known for providing farmers with the tools they need is offering to help them survive and find new revenue streams. That's value innovation!

8. Intangible Value

Most companies are worth more than their hard assets, and in some cases much more. Brand value—or corporate reputation, more generally—can be worth many billions of dollars. Any threat to that value has to be taken seriously. From BP to GE to Wal-Mart, a growing number of companies have launched campaigns to build a green element into their brand.

THE HAT TRICK GREEN-TO-GOLD PLAY AT ALCAN

We've set up these Green-to-Gold Plays as if they were distinct strategies. That's the easiest way to think about them and find business opportunities. But nothing says a company can't do everything—lower costs and risk *and* drive revenues—at the same time.

Alcan, the $20-billion Canadian aluminum and packaging company, recently pulled off this impressive feat. Quick background: Producing aluminum is dirty work with large-scale environmental consequences. The industry is one of most energy-intensive in the world, and mining and smelting produce large quantities of waste, including what settles at the bottom of the big pots they use in production. This toxic residue—not surprisingly called potlining—is scraped off, which is where the environmental challenge comes in.

Nobody really knows what to do with this toxic slurry since it's not easy to recycle. Mostly it just sits around, taking up space and creating liabilities. Alcan's Dan Gagnier estimates that just his company's backlog of potlining is over half a million tons. But that's about to change. Alcan has developed an innovative new technology to turn potlinings into inert, recycled material. The company is investing $150 million in a prototype treatment facility. This breakthrough solves Alcan's waste problem and may even leave the company treating its *competitors'* waste—for a price, of course.

As CEO Travis Engen says, "We wouldn't have come to this without the framework of sustainability." Alcan's initial goal was to reduce an environmental burden. But executives are anticipating that the new process will end up lowering waste, reducing risk and liability, and generating revenues. Alcan is building a real Eco-Advantage.

HURDLES

Executing a corporate environmental strategy is never easy. A range of hurdles can trip up even the most sophisticated company, making the quest for Eco-Advantage elusive. We've identified thirteen primary sources of failure in environmental strategy (see table):

First we've identified failures of planning, where the focus of the environmental initiative isn't well thought out or expectations are out of whack. Second, we see failures of organization, including competing demands placed on middle management and silo thinking,

Eco-Advantage Hurdles

Failures of . . .	
Planning	• Seeing the trees but not the forest • Misunderstanding the market • Expecting a price premium • Misunderstanding customers • Seeking perfection • Ignoring stakeholders
Organization	• Middle-management squeeze • Silo thinking • Eco-isolation of environmental professionals • Inertia
Substance	• Claims outpacing actions • Surprises and unexpected consequences • Failing to tell the story

which limits coordination and chills creativity. No wonder inertia is so often the by-product of organizational weakness. Third, failures of substance are wide-spread. Sometimes the problem can be traced to not leveraging environmental gains. In other cases, poor results come from trying to leverage claims that have too much hot air. All of these failings are common but can be overcome by building an Eco-Advantage framework and by equipping employees up and down the line with the right tools.

TYING THE FRAMEWORK TOGETHER

There's no single right way to assemble all the pieces of an Eco-Advantage strategy. Some of the elements in the framework we've provided will fit an organization perfectly, while others may not. Some are better than others at any given point in a product or company's evolution. While each has an independent logic, the major elements do work together.

The plays and tools are integrally linked and feed one another. To see how they fit together, we've developed a summary table, which is found in Appendix III. Collectively, these plays and tools reinforce each other to lower business risk and increase business value. Exploring a company's footprint with a Life Cycle Assessment, for example, will identify opportunities to cut waste and cost, which translates directly into lower business risk. Looking downstream and identifying ways to lower a customer's environmental burden also reduces everyone's risk and improves performance along the value chain. And identifying new market spaces creates added value and diminishes the chance of the market shifting out from under you.

GREEN TO GOLD

Eco-Advantage has a twin logic. On one hand, the strategic gains we've identified are based on hard-edged analysis. In a world of constrained natural resources and pollution pressures, the business case for environmental stewardship grows stronger every day. Pressures on companies now come not only from screaming eco-radicals, but also from traditional "white-shoe" bankers and others asking tough-minded questions about environmental risk and liability. Those who offer solutions to society's environmental problems both mute their potential critics and find expanding markets. As Timberland's CEO, Jeff Swartz, said recently about one of his company's environmental initiatives, "I can now make the fact-based case to the hardest-nosed engineer in the world. . . . That's not limousine liberal, not self-indulgent. It's hard-nosed business. That is the innovation we seek."

In parallel, there's a strong values component to the case for corporate environmental care. The WaveRiders we've studied have made money—lots of it—by refining their business strategies to incorporate environmental factors. But as much as they are driven by profits, they are also aware that their stewardship helps more than the bottom line. When short-term gains don't justify green initiatives, they are willing to look for long-term value for themselves and their workers, for their communities, and for the planet. The gold they've discovered by going green is not only about money.

More and more people in the business world see corporations playing a major role in solving the world's environmental problems. Business, they know, is our most powerful mechanism for creating a functioning society and matching needs with goods and services.

Companies can and should be a force for good, leading the charge on caring for the environment and protecting our shared natural assets. Financial *and* environmental success can be achieved together. With the right mindset and tools, companies can handle the hard trade-offs.

New values-centered executives are creating companies that inspire employees and customers alike. In the end, Eco-Advantage is about a new way for inspired people—executives, managers, and workers—to build companies and industries that are not just innovative, powerful, and great . . . but good too.

Appendix I: Additional Resources

Plenty of help is available for companies looking to fold environmental considerations into their business strategies. The books and other resources listed here are a starting point. Further resources, including business school case studies and a list of commonly used acronyms, can be found on our website: www.eco-advantage.com.

Selected Green Business Books

Anderson, Ray. *Mid-Course Correction: Toward a Sustainable Enterprise: The Interface Model*. White River Junction, VT: Chelsea Green Publishing, 1998.

Bendell, Jem, editor. *Terms for Endearment*. Sheffield, UK: Greenleaf Publishing, 2000.

Benyus, Janine. *Biomimicry: Innovation Inspired by Nature*. New York: HarperCollins, 1997.

Cairncross, Frances. *Costing the Earth*. London: Economist Books, 1992.

Elkington, John. *Cannibals with Forks: The Triple Bottom Line of 21st Century Business*. Oxford: Capstone Publishing, 1997.

———. *The Chrysalis Economy: How Citizen CEOs and Corporations Can Fuse Values and Value Creation*. Oxford: Capstone Publishing, 2001.

Elkington, John, and Julia Hailes. *The Green Consumer Guide*. London: Gollancz, 1988 (U.S. edition with co-author Joel Makower. New York: Penguin, 1990).

Epstein, Marc J. *Measuring Corporate Environmental Performance: Best Practices for Costing and Managing an Effective Environmental Strategy*. Burr Ridge, IL: Institute for Management Accounting and Irwin Professional Publishing, 1996.

Epstein, Marc J., and B. Birchard. *Counting What Counts: Turning Corporate Accountability Into Competitive Advantage*. Reading, MA: Perseus Books, 1999.

Gunningham, Neil A., Robert A. Kagan, and Dorothy Thornton. *Shades of Green: Business, Regulation, and Environment*. Palo Alto, CA: Stanford University Press, 2003.

Hart, Stuart L. *Capitalism at the Crossroads: The Unlimited Business Opportunities in Solving the World's Most Difficult Problems*. Upper Saddle River, NJ: Wharton Publishing School, 2005.

Hawken, Paul. *Ecology of Commerce: A Declaration of Sustainability*. New York: HarperCollins, 1993.

Hawken, Paul, Amory Lovins, and L. Hunter Lovins. *Natural Capitalism: Creating the Next Industrial Revolution*. Boston: Back Bay Books, 1999.

Hoffman, Andrew J. *From Heresy to Dogma: An Institutional History of Corporate Environmentalism*. Stanford, CA: Stanford University Press, 2001.

Holliday, Charles O., Jr., Stephan Schmidheiny, and Philip Watts. *Walking the Talk: The Business Case for Sustainable Development*. Sheffield, UK: Greenleaf Publishing, 2002.

McDonough, Bill, and Michael Braungart. *Cradle to Cradle: Remaking the Way We Make Things*. New York: North Point Press, 2002.

Prakash, Aseem. *Greening the Firm*. Cambridge, UK: Cambridge University Press, 2000.

Reinhardt, Forest. *Down to Earth: Applying Business Principles to Environmental Management*. Cambridge, MA: Harvard Business School Press, 2000.

Schmidheiny, Stephen, with the Business Council for Sustainable Development. *Changing Course: A Global Business Perspective on Development and the Environment*. Cambridge, MA: MIT Press, 1992.

Schmidheiny, S., F. J. Zorraquin, and World Business Council for Sustainable Development. *Financing Change: The Financial Community, Eco-Efficiency, and Sustainable Development*. Cambridge, MA: MIT Press, 1996.

Taylor, J. Gary, and Patricia Scharlin. *Smart Alliance: How a Global Corporation and Environmental Activists Transformed a Tarnished Brand*. New Haven, CT: Yale University Press, 2004.

Vogel, David. *The Market for Virtue: The Potential and Limits of Corporate Social Responsibility*. Washington, D.C.: The Brookings Institute, 2005.

von Weizacker, Ernst, Amory B. Lovins, and L. Hunter Lovins. *Factor Four: Doubling Wealth—Halving Resource Use: A Report to the Club of Rome*. London: Earthscan Publications, 1998.

Winsemius, Peter, and Ulrich Guntram. *A Thousand Shades of Green: Sustainable Strategies for Competitive Advantage*. London: Earthscan Publications, 2002.

Selected Green Business Articles

Gladwin, Thomas N. "Environmental Policy Trends Facing Multinationals." *California Management Review* 20, no.2 (1977): 81–93.

Hart, Stuart L. "Beyond Greening: Strategies for a Sustainable World." *Harvard Business Review* 75, no.1 (1997): 66–76.

Hoffman, Andrew J. "Climate Change Strategy: The Business Logic Behind Voluntary Greenhouse Gas Reductions." *California Management Review* 47, no.3 (2005): 21–46.

Lovins, Amory B., L. Hunter Lovins, and Paul Hawken. "A Road Map for Natural Capitalism." *Harvard Business Review* 77, no.3 (1999): 145–159.

Packard, Kimberly O'Neill, and Forest Reinhardt. "What Every Executive Needs to Know About Global Warming." *Harvard Business Review* 78, no.4 (2000): 129–135.

Porter, Michael. "America's Green Strategy." *Scientific American* 264 (1991): 168.

Porter, Michael E., and Claas Van Der Linde. "Green and Competitive: Ending the Stalemate." *Harvard Business Review* 73, no.5 (1995): 120–134.

Reinhardt, Forest L. "Bringing the Environment Down to Earth." *Harvard Business Review* 77, no.4 (1999): 149.

Repetto, Robert, and Duncan Austin. "An Analytical Tool for Managing Environmental Risks Strategically." *Corporate Environmental Strategy* 7, no.1 (2000): 72–84.

Steger, Ulrich. "Corporations Capitalize on Environmentalism." *Business and Society Review* 75, no.3 (1990): 72–73.

Thornton, Dorothy, Robert A. Kagan, and Neil Gunningham. "Sources of Corporate Environmental Performance." *California Management Review* 46, no.1 (2003): 127–141.

Vogel, David J. "Is There a Market for Virtue? The Business Case for Corporate Social Responsibility." *California Management Review* 47, no.4 (2005): 19–45.

———. "The Low Value of Virtue." *Harvard Business Review* 83, no.6 (2005).

Walley, Noah, and Bradley Whitehead. "It's Not Easy Being Green." *Harvard Business Review* 72, no.3 (1994): 46–51.

Environment-Oriented Magazines

Audubon

Conservation International e-news updates (at www.conservation.org)

Ethical Corporation

Friends of the Earth eNews (at www.foe.org)

Green Futures (UK)

Green@Work

E/The Environmental Magazine

Nature Conservancy

On Earth (Natural Resources Defense Council)

Rainforest Alliance newsletter (at www.ra.org)

Sierra Magazine
This Green Life (quarterly newsletter from Natural Resources Defense Council)
World Wildlife Fund newsletter (at www.wwf.org)

Websites and Blogs with an Environmental Focus
www.commonsblog.org (free-market environmentalism)
www.csrwire.com
www.eco-advantage.com
www.enn.com/business_news_main_d.html (Environmental News Network)
www.env-econ.net (Environmental Economics)
http://feeds.feedburner.com/greenthinkers
www.gemi.org
www.greenbiz.com and www.climatebiz.com
www.gri.org
http://gristmill.grist.org
www.makower.typepad.com/joel_makower
www.sustainability.com
www.sustainabilitydictionary.com
www.sustainablebusiness.com
www.sustainablemarketing.com
www.sustainablog.blogspot.com
www.treehugger.com
www.triplepundit.com
www.wbcsd.org
www.worldchanging.com

Appendix II: Methodological Overview

We began our research in 2003 by asking a seemingly simple question: Which companies are environmental leaders? Finding a financial leader is fairly straightforward. Look for excellent performance on well-known metrics such as net income or cash flow, or just graph the stock price over time. In the environmental realm, however, picking a leader is not as clear. In an ideal world, every company would release precise and comparable metrics on environmental outcomes (emissions, permit violations, and so on). But that kind of data is just not available. So we gathered the available information from a variety of sources, including public records and the academic literature, and reviewed the rankings that do exist. We also surveyed environmental executives, asking them which companies have leading edge environmental positions.

Our rankings include some outcome-based metrics, but much of what is available is survey-based and anecdotal. Indeed, most of the environment and sustainability rating services—such as New York–based Innovest, Zurich-based Sustainable Asset Management, and others in the socially responsible investment community—rely heavily upon company questionnaires.

By combining existing rankings and some additional data, we created our own indicative meta-ranking. Our scoring is based on:

I. SRI Rankings (25 Percent)

- Major indices (80 percent): Innovest AAA or AA rating, Domini 400 (com-

piled by KLD Research and Analytics), FTSE4Good Global, Dow Jones Sustainability Index.
- Additional indices (20 percent): Calvert Funds (Top 10 holdings), Sustainable Business All-Star 20, Sierra Club Mutual Fund.

II. Engagement in Compacts and/or Agreements (25 Percent)

- Reporting or Corporate Social Responsibility (33 percent): Ceres, Global Reporting Initiative, Global Environmental Management Initiative, Global Compact, Business for Social Responsibility, World Business Council for Sustainable Development.
- Climate Change (33 percent): Pew Center's Business Environmental Leadership Council, EPA Climate Leaders, Department of Energy Voluntary Report of GHG emissions, Climate Savers (World Wildlife Fund), Partners for Climate Action, Chicago Climate Exchange, Other/Regional agreements.
- EPA Programs (33 percent): Energy Star, National Performance Track.

III. Surveys (25 Percent)

- Green-to-Gold survey of environmental executives (see below) (75 percent).
- Other surveys (25 percent): Center for Environmental Innovation survey of EHS executives, *Financial Times* most respected survey of CEOs and NGO leaders.

IV. Other (25 Percent)

- Review of literature of green business (33 percent): Frequency of coverage.
- Outcomes (33 percent): Performance on eight key metrics: energy use, water use, hazardous waste, total waste, and emissions of greenhouse gases, volatile organic compounds, nitrogen oxides (NOx), and sulfur oxides (SOx).
- Lobbying/Political behavior (33 percent): Donations to U.S. Congress, weighted by representative's ranking by the League of Conservation Voters.

Our Survey

Defining an environmental leader is challenging. A core component of any ranking should be based on the opinions of knowledgeable people from the business and regulatory communities. We asked a sampling of environmental executives—plus a group from the U.S. EPA—a few simple questions, including:

- What companies do you see as leaders in environmental performance and strategy?

- Are there any companies which you think have received positive attention as environmental leaders but have not actually earned the praise?
- Are there any companies which you think have been unfairly criticized for their environmental actions and which have, in fact, strong environmental performance and strategies?

These were open-ended questions, so the answers were completely driven by the knowledge and perceptions of the executives we surveyed. The top vote-getters on these three questions—which we call "Environmental Leaders," "Overrated," and "Underrated," respectively—are listed in the table, sorted from most responses to least.

For the most part, the results were unsurprising and reinforced the other sources of data. The environmental leaders list closely mirrors our final WaveRiders list in Chapter 1 (of course, this survey represented only about 19 percent of the weighting in the total ranking—75 percent of the 25 percent weighting in category III above—so the list is not precisely the same). But we also notice some interesting findings. A number of companies—like DuPont, Shell, and Interface—are on all three lists. What does this mean? In part, it indicates that defining a leader is in the eye of the beholder. It also suggests that no company is a leader in all dimensions of environmental strategy and may lag in some areas. Some survey respondents will focus on the lagging and others on the leading.

The Underrated list may get beneath the surface of public perception. GE's strong showing, for example, reflects the fact that many EHS professionals recognize that the company had a cutting-edge environmental management system even while Jack Welch battled the EPA, harming the public perception of the company's environmental posture.

How did favorite environmental punching bag ExxonMobil end up with the most Underrated votes? This result reflects the knowledge among corporate environmental officials that the company runs an efficient operation, and lower waste translates into lower levels of emissions. Moreover, ExxonMobil is seen by many to have a top-tier environmental management system. Even so, no respondent picked the company as a leader outright.

Data Problems and Biases in the Ranking

Rankings entail many challenges and limitations. Assumptions and short-cuts to overcome data gaps cannot be avoided. So our list should not be seen as a *definitive* ranking of corporate environmental leaders. Some additional observations and caveats should be noted:

- The ranking emphasizes—more than we would like—public perceptions and environmental efforts, not results or outcomes. Companies with a high profile among peers and rating agencies tend to stand out. Lower-profile companies that go about their business quietly, like Herman Miller, not to mention privately held companies, like IKEA, don't score as highly. In

Green-to-Gold Survey Results—Sorted by Number of Responses

Environmental Leaders	Overrated	Underrated
BP	Ford	ExxonMobil
Dow	BP	Nike
DuPont	Home Depot	Shell
Interface	3M	Starbucks
Toyota	Shell	BP
Shell	Interface	McDonald's
J&J	McDonald's	GE
IBM	Monsanto	DuPont
Nike	Nike	Department of Defense
3M	GM	Monsanto
HP	Shaw	Alcoa
Patagonia	Knoll	BNSF
Unilever	The Body Shop	Union Pacific
Ford	Ontario Power	CSX
SC Johnson	ChevronTexaco	Staples
Novo Nordisk	Chiquita	Citigroup
P&G	MDBC	ChevronTexaco
Baxter	Patagonia	Toyota
Bristol-Myers Squibb	Weyerhaeuser	Interface
Anheuser-Busch	Lucent	Bayer
Aveda	Xerox	Dow
BASF	Compaq	Weyerhaeuser
Herman Miller	Apple	Dell
Norsk Hydro	ArcherDaniels	GM
Ricoh	Dow	Koch Industries
Stonyfield Farms	Levi Strauss	Ford
Electrolux	DuPont	UPM-Kymmene
Alcoa	Volkswagen	
Kinko's	STMicroelecronics	
Sony	Teknion	
Xerox	BC Hydro	
Starbucks	Canfor	
Ben & Jerry's	Alcan	
Cargill-Dow	Alcoa	
Cooperative Bank	Bristol-Myers Squibb	
GE	ExxonMobil	
Henkel	Rio Tinto	
Honda	Lafarge	
IKEA	IBM	
Intel	Siemens	
JM	Fujikura Kasei	
MDBC	Vattenfall	
Milliken		
Motorola		
Mtn Equip Coop		
Norm Thompson		
Novartis		
Philips		
Pitney Bowes		
Suncor		
Swiss Re		
UPS		
H&M		

fact, while IKEA ranks 24th on our international WaveRiders list on page 25, it's arguably the most sustainable large company in the world.

- With regard to private companies, we "imputed" a score on the SRI dimension equal to the average SRI score of the other top companies.
- Recognizing that our own analysis—and all others—probably puts too much weight on visible evidence of commitment to environmental stewardship rather than actual on-the-ground environmental performance, we made a special effort to uncover underrated companies and high flyers with low profiles that might have interesting stories to tell.
- Important data are not available or are shaky. In particular, data on actual environmental outcomes—like emissions or energy use—are not comparable across companies. In a world with consistently collected data and comparable metrics, the rankings could be based on actual performance rather than measure of effort and perception.
- The SRI community bases much of its ranking on company surveys. Relying on self-reported information leads to a series of flaws and biases that have been well-documented in the research methodology literature.
- Some duplication exists in the factors analyzed. We include many of the same elements as the SRI community, so the scores are self-reinforcing. Those that get good rankings by Innovest, for example, will likely score well on other dimensions of our ranking.

We believe that, while primitive in a number of ways, a "ranking of rankings" such as ours produces useful results. By combining multiple data sources and scoring systems, some of the noise can be eliminated and a clearer "signal" emerges. The rankings, of course, only provided a starting point for our interviewing and analysis.

Appendix III: Most Relevant Tools for Each Green-to-Gold Play

In this table we attempt to align the major concepts of the book. For each Green-to-Gold play, we suggest the mindset principles and tools that would best enable that strategy.

Relevant Principles and Tools for Each Green-to-Gold Play

Green-to-Gold Play	Mindset Principles	Eco-Tracking Tools	Redesign Tools	Culture-Building Tools
1. Eco-efficiency *and* 2. Eco-expense reduction	— Broad thinking (payoffs) — Apollo 13 — Do the right thing	— Life Cycle Assessment — Data/metrics — Systems	— Design: DfE, closed loops	— Stretch goals — Decisions: hurdle rate, pairing, internal pollution "markets" — Ownership: incentives, awards — Storytelling: training
3. Value chain eco-efficiency	— Broad thinking (boundaries or full value chain)	— Trace footprint: LCA	— Design: DfE, closed loops	— Stretch goals — Storytelling: training
4. Eco-risk control	— Broad thinking (time, payoffs, boundaries) — Feelings are facts	— Scenarios — Data/metrics: materials database — Management systems: emergency procedures — Partnering	— Supply chain audits	— Storytelling: training — Awards
5. Eco-design	— Apollo 13 (meet customer needs *and* environmental goals)	— Trace footprint: LCA	— Design: DfE	— Stretch goals — Ownership: incentives, awards, cross-fertilization

Green-to-Gold Play	Mindset Principles	Eco-Tracking Tools	Redesign Tools	Culture-Building Tools
6. Eco-sales and marketing	— Feelings are facts (understand customers)	— Data/metrics (for justifiable ad claims)		— Storytelling
7. Eco-defined new market space	— Broad thinking about payoffs — CEO commitment (high-risk investments) — Do the right thing	— Trace footprint: LCA — Scenarios	— Design: DfE	— Decisions: hurdle rate — Ownership: incentives, awards
8. Intangible value	— Broad thinking (payoffs) — CEO commitment — Do the right thing	— Partnering: NGOs, communities	— Green building	— Ownership: incentives, steward roles, mission review — Storytelling: CSR reports

Notes

INTRODUCTION. THE ENVIRONMENTAL LENS

p. 1, PlayStation game systems: Details on the PlayStation situation from Teruo Masaki (Sony), "Keynote II" (speech, Business for Social Responsibility Conference, Los Angeles, CA, 11 November 2003); see also Associated Press, "Dutch Authorities Stop Sony Game Console Shipment over Environmental Fears," 4 December 2001.

PART ONE. PREPARING FOR A NEW WORLD

CHAPTER 1. ECO-ADVANTAGE

p. 7, Immelt announces a new initiative, "ecomagination": Jeff Immelt, Speech, launch event for ecomagination, Washington, D.C., 9 May 2005.

p. 7–8, Wal-Mart CEO Lee Scott: Lee Scott, "Twenty-First Century Leadership," Speech, made to Wal-Mart shareholders, 24 October 2005, http://walmartstores.com/Files/21st%20Century%20Leadership.pdf.

p. 8, "a green revolution": David Ignatius, "Corporate Green," *Washington Post*, 11 May 2005, A7.

p. 9, $6 billion of new business: Joel Makower, "The State of Green Business," 31 December 2005, http://makower.typepad.com/joel_makower/2005/12/the_state_of_gr.html; see also HP Citizenship Report, 2005, at www.hp.com/hpinfo/globalcitizenship/gcreport/products/dfe.html.

p. 9, Goldman Sachs announced: Claudia Deutsch, "Goldman to Encourage Solutions to Environmental Issues," *New York Times*, 22 November 2005.

p. 9, signed on to the "Equator Principles": See Benjamin C. Esty, "The Equator Principles: An Industry Approach to Managing Environmental and Social Risks," Harvard Business School Case N9-205-11, 2005.

p. 11, traditional elements of competitive advantage: Michael Porter notes in *Competitive Advantage of Nations* (New York: Free Press, 1990) that the modern competitive edge does not come primarily from low-cost inputs, but rather from innovation. Thomas Friedman makes a similar argument in *The World Is Flat* (New York: Farrar, Straus and Giroux, 2005).

p. 12, "We have two maxims": Phil Berry, interview by author, Toronto, Canada, 10 March 2006.

p. 12, "prisoner of the market": Albert Bressand (Shell), interview by author, London, England, 7 May 2004.

p. 12, 1984 disaster in Bhopal, India: D. Kurzman, *A Killing Wind: Inside Union Carbide and the Bhopal Catastrophe* (New York: McGraw-Hill, 1987).

p. 13, protect the company's "license to grow": Don Moseley (Wal-Mart), interview by author, New Haven, CT, 24 October 2004.

p. 14, "twenty years to build a reputation": This quote is commonly attributed to Warren Buffett; see, for example, Anne Fisher, "America's Most Admired Companies," *Fortune*, 6 March 2006.

p. 14, Milton Friedman: Milton Friedman, "The Social Responsibility of Business is to Increase Its Profits," *New York Times*, 13 September 1970 (Magazine at 33).

p. 15, "flattening" of the global markets: Friedman, *The World Is Flat*.

p. 15, willingness to pay a premium: See "2005 Yale Environment Poll," Yale Center for Environmental Law and Policy, 2005, www.yale.edu/envirocenter.

p. 16, regulatory system in America: David Vogel, "Comparing Environmental Governance: Risk Regulation in the EU and the US," Center for Responsible Business, Working Paper Series, Paper 2, 1 September 2003, http://repositories.cdlib.org/crb/wps/2.

p. 16, cars in China and India: James Brooke, "At Tokyo Auto Show, a Focus on Fuel, Not Fenders," *New York Times*, 4 November 2005, C1, 4.

p. 17, "companies cannot succeed": see this quote in Monica Araya, "To Tell or Not to Tell? Determinants of Environmental Disclosure and Reporting in Corporate Latin America" (Ph.D. dissertation, Yale University, 2006), 276.

p. 18, the Internet "has given the angry voices": Claudia Deutsch, "Take Your Best Shot; New Surveys Show That Big Business Has a P.R. Problem," *New York Times*, 9 December 2005.

p. 22, environmental challenges alongside social issues: See Daniel Esty, "A Term's Limits," *Foreign Policy* (September–October 2001): 74–75; for more on the bottom of the pyramid, see C. K. Prahalad, *The Fortune at the Bot-*

tom of the Pyramid (Upper Saddle River, NJ: Wharton School Publishing, 2004).

p. 22, *business case* for taking up the social agenda: David Vogel, *The Market for Virtue: The Potential and Limits of Corporate Social Responsibility* (Washington, D.C.: Brookings Institution Press, 2005).

p. 24, lacks the structure and rigor provided in the financial realm: A range of organizations are trying to establish environmental performance metrics for companies, including the Global Reporting Initiative (www.globalreporting.org), the Global Environmental Management Initiative (www.gemi.org), and the World Business Council on Sustainable Development (www.wbcsd.org). WBCSD, for example, has produced a 2005 report titled "Beyond Reporting: Creating Business Value and Accountability."

p. 24, environmental and sustainability scorecards: Innovest is a stock advisory service based in New York (see www.innovest.com); SAM is a service based in Zurich (see www.sam-group.com). For a complete list of organizations included in our ranking, see Appendix II.

p. 26, stock performance of the publicly held WaveRiders: Four of the WaveRiders are privately held (SC Johnson, Ben & Jerry's, Patagonia, and IKEA), and historical stock prices were not available for another three (Holcim, Henkel, Novozymes). The figure tracks the stock price movement of the other forty-three.

p. 26, correlation is not causation: Many researchers and studies have attempted to demonstrate a connection between environmental and financial performance. See Andrew King and Michael Lenox, "Does It Really Pay to Be Green? An Empirical Study of Firm Environmental and Financial Performance," *Journal of Industrial Ecology* 5, no. 1 (2001): 105–116; M. Orlitzsky, F. Schmidt, and S. Rynes, "Corporate Social and Financial Performance: A Meta-Analysis," Social Investment Forum Foundation, 2004, www.socialinvest.org/Areas/Research/Moskowitz/winning_papers.htm; K. Gluck and Y. Becker, "The Impact of Eco-Efficiency Alphas on an Actively Managed U.S. Equity Portfolio Performance," State Street Global Advisors, February 2004; see also Vogel, *The Market for Virtue*.

p. 28–29, "What would it take for Wal-Mart": Scott, "Twenty-First Century Leadership."

p. 29, "To be a great company": Immelt, ecomagination speech.

CHAPTER 2. NATURAL DRIVERS OF THE GREEN WAVE

p. 30, 100 percent of its fish from sustainable sources: Unilever has fallen short of this ambitious goal, but has made significant strides. We discuss this in more detail in chapter 10. Note also that in 2006, Unilever announced plans to sell most, but not all, of its fish business.

p. 31, "it is in Unilever's commercial interest": "Unilever acts to improve sustainability of fishing in European waters," *Business Wire*, 22 April 1996.

p. 31, "we will be out of business": S. Sanandakumar, "Unilever to Buy Seafood Only from Certified COs," *The Economic Times*, 11 February 2003.

p. 31, Globally, though, the trend lines are broadly negative: For a fact-based review of the world scene on a country-by-country and issue-by-issue basis, see Yale Center for Environmental Law and Policy, "2006 Environmental Performance Index," 2006, http://www.yale.edu/epi.

p. 31, growing dangers of climate change: Eugene Linden, "Climate Shock," *Fortune*, 23 January 2006, 135–145.

p. 32, assets on the planetary balance sheet: See Paul Hawken, Amory Lovins, and L. Hunter Lovins, *Natural Capitalism: Creating the Next Industrial Revolution* (Boston: Back Bay Books, 1999).

p. 32, "you really have only two assets": Mats Lederhausen, conversation with author, 19 August 2004.

p. 34, over the last 650,000 years: Andrew C. Revkin, "Rise in Gases Unmatched by a History in Ancient Ice," *New York Times*, 25 November 2005.

p. 34–35, over 900 peer-reviewed articles: Naomi Oreskes, "Beyond the Ivory Tower: The Scientific Consensus on Climate Change," *Science* 306 (2004): 1686.

p. 35, where greenhouse gases come from: Pew Center on Global Climate Change, "Global Warming Basics: Facts and Figures," www.pewclimate.org/global-warming-basics/facts_and_figures/.

p. 36, heat wave in Europe: James Reynolds, "Urgent Action Is Called for Now, Says Blair in Dire Global Warning," *The Scotsman*, 15 September 2004.

p. 36, half the summers: Juliet Eilperin, "Humans May Double the Risk of Heat Waves," *Washington Post* (Final Ed.), 2 December 2004.

p. 36, ice is melting all over the world: Bill Chameides and James Wang, Global Warming's Increasingly Visible Impacts, Environmental Defense, 2005, http://www.environmentaldefense.org/documents/4892_GlobalWarming mpacts_ExSummary.pdf; David Laskin, "The Great Meltdown: The Freezing Cold Stuff Is a Hot Topic for Scientists and Adventurers Alike," review of "Thin Ice: Unlocking the Secrets of Climate in the World's Highest Mountains by Mark Bowen," *Washington Post*, 20 November 2005.

p. 37, pine bark beetle: Associated Press, "Canadian Beetle Infestation Worries U.S.," 16 January 2006.

p. 37, U.S. Midwest may become as drought plagued: Amy Royden, "U.S. Climate Change Policy Under President Clinton: A Look Back," in *Rio's Decade: Reassessing the 1992 Earth Summit, Golden Gate U.L. Rev* 32 (2002): 415 n.289.

p. 37, tragic scenes from post-Katrina New Orleans: Some have made the argument that the damage in New Orleans was a direct result of climate change strengthening storms. See Ross Gelbspan, "Katrina's Real Name," *Boston Globe*, 30 August 2005.

p. 38, The economic cost of extreme natural disasters: World Water Council,

Number of Killer Storms and Droughts Increasing Worldwide, 27 February 2003 (Press Release), http://www.worldwatercouncil.org/fileadmin/wwc/News/WWC_News/News_2003/PR_climate_27.02.03.pdf.

p. 38, **summit of power company leaders:** Adam Aston and Burt Helm, "The Race Against Climate Change," *BusinessWeek*, 12 December 2005, 59.

p. 39, **the era of "Hydrocarbon Man":** see Daniel Yergin, *The Prize: The Epic Quest for Oil, Money & Power* (New York: Free Press, 1991). For predictions of the end of the oil era see Colin J. Campbell and Jean H. Laherrere, "The End of Cheap Oil," *Scientific American*, March 1998, 78–83; and Matthew R. Simmons, *Twilight in the Desert: The Coming Saudi Oil Shock and the World Economy* (Hoboken, NJ: Wiley, 2005).

p. 40, **two-thirds of U.S. electricity:** U.S. fuel mix statistics from Edison Electric Institute, see http://www.eei.org/industry_issues/industry_overview_and_statistics/industry_statistics/index.htm.

p. 40, **support for nuclear power:** Peter Schwartz and Spencer Reiss, "Nuclear Now! How Clean, Green Atomic Energy Can Stop Global Warming," *Wired*, February 2005; also, in 2006 former EPA Administrator Christie Todd Whitman and Greenpeace co-founder Patrick Moore announced they were leading a public relations campaign for new nuclear reactors: see Matthew L. Wald, "Ex-environmental Leaders Tout Nuclear Energy," *New York Times*, 25 April 2006, A24.

p. 41, **clean tech funds:** Data from the chart on clean tech is from N. Parker and A. O'Rourke, *The Cleantech Venture Capital Report 2006*, The Cleantech Venture Network, 2006. Size of clean tech markets is from "Trillions Foreseen for Business in Global Environmental Markets," *The Environment Forum*, 29 November 2005, cited from the Environmental News Service.

p. 43, **goes untreated directly into rivers:** Millennium Ecosystem Assessment, "Ecosystems and Human Well-Being: Synthesis," (Washington, D.C.: Island Press, 2005), http://www.millenniumassessment.org/en/products.aspx.

p. 43, **problems of water quantity . . . in China:** James Kynge, "China Counts Economic Costs as Water Shortages Hamper Growth," *Financial Times*, 26 March 2004.

p. 43, **34,000 Chinook and Coho salmon died:** Glen Martin, "Salmon Kill Linked to Level of Klamath River's Flow—Reduced for Irrigation—Played a Role in Huge Die-off, U.S. Study Finds," *San Francisco Chronicle*, 19 November 2003.

p. 44–45, **Valvidia pulp and paper mill:** Monica Araya, "To Tell or Not to Tell? Determinants of Environmental Disclosure and Reporting in Corporate Latin America," (Ph.D. Dissertation, Yale University, 2006), 286–292.

p. 45, **"solve the world's most pressing water . . . needs":** GE Infrastructure, "GE Technology Plays Critical Role in Opening of World's Largest Potable Ultrafiltration Plant," 6 September 2005 (Press Release), http://www.gewater.com/pdf/pr/20050906_pr.pdf.

p. 45, **"Our personal health":** From Rio Tinto internal document, "Sustaining a Natural Balance: A Practical Guide to Integrating Biodiversity into Rio Tinto's Operational Activities," provided by Dave Richards (Rio Tinto).

p. 45, value of biodiversity: R. Costanza et al., "The Value of the World's Ecosystem Services and Natural Capital," *Nature*, 387 (1997): 253–260.

p. 46, sixth major extinction period: see, for example, Edward O. Wilson, *The Future of Life* (New York: Alfred A. Knopf, 2002), 99.

p. 46, extinction of many frog species: Jeffrey Kluger, "Why Are These Frogs Croaking?," *Time*, 23 January 2006.

p. 46, pesky mollusks cost the power industry alone $3.1 billion: see www.dgif.state.va.us/zebramussels/.

p. 47, preserve land . . . 130,000 acres: Wal-Mart "Conservation," See http://walmartstores.com/GlobalWMStoresWeb/navigate.do?catg=443.

p. 47, stunning products for human technology to copy . . . spider silk: Janine Benyus, *Biomimicry* (New York: William Morrow and Company, 1997).

p. 48, lands in the Catskill Mountains: Daniel Hirsch, " 'Wetlands' Importance Now Made Clear," *Atlanta Journal-Constitution*, 12 September 2005, Home Edition, 11A.

p. 48, Perrier subsidizes: "Are You Being Served? Environmental Economics," *The Economist*, 23 April 2005.

p. 49, driven into bankruptcy: Mark D. Taylor, "How Congress Can Solve the Great Asbestos Bankruptcy Heist," *Mealey's Asbestos Bankruptcy Report*, April 2005, 1–10.

p. 49, Endocrine disruptors: Natural Resources Defense Council, "Endocrine Disruption: An Overview and Resource List," September 1998, www.nrdc.org/health/effects/bendrep.asp.

p. 49, the annual number of articles: Dr. Taisen Iguchi (National Institute for Basic Biology, Japan), personal communication with author, 14 February 2006.

p. 50, ban PBDEs by 2008: Jennifer Lee, "California to Ban Chemicals Used as Flame Retardants," *New York Times*, 10 August 2003, 14; Renee Sharp and Sonya Lunder, "In the Dust: Toxic Fire Retardants in American Homes," Environmental Working Group, 2004, http://www.ewg.org/reports_content/inthedust/pdf/inthedust_final.pdf.

p. 50, prenatal mercury exposure: Mercury study referenced in Net Impact CSR weekly newsletter, #100 (www.netimpact.org).

p. 50, Roughly 10 to 20 percent of women: Juliet Eilperin, "Excess Mercury Levels Increasing; Survey Shows Fifth of Women of Childbearing Age Are Affected," *Washington Post*, 21 October 2004, A2.

p. 51, Even WaveRiders Can Take a Tumble: Lisa Roner, "U.S. News: Environment Agency Says DuPont Withheld Teflon's Chemical Danger," *Ethicalcorporation.com*, 13 July 2004. Also Michael Janofsky, "DuPont to Pay $16.5 Million for Unreported Risks," *New York Times*, 15 December 2005, Late Edition–Final; and Dawn Rittenhouse, Jayne Seebach, Daniel Taylor (DuPont), personal communication with author, 13 March 2006.

p. 52, concentration of sulfur dioxide: US EPA, "More Details on Sulfur Dioxide—Based on Data Through 2002," http://www.epa.gov/airtrends/sulfur2.html.

p. 52, air pollution causes over 300,000 premature deaths: European Commission, "Brief Commission Prepares Strategy to Improve Air Quality," 22 February 2005, see www.eurativ.com/en/health.

p. 53: Lafarge . . . led its industry: World Business Council for Sustainable Development, *The Cement Sustainability Initiative Progress Report*, June 2005.

p. 53, quarter of a million dollars: Larry Dykhuis, Randy Ruster, Kent Gawort (Herman Miller), interview by author, Zeeland, MI, 4 May 2004.

p. 53, the United States recycles: For more details on U.S. recycling rates, see http://www.epa.gov/epaoswer/non-hw/muncpl/recycle.htm#Figures. International comparisons from the "Recycling Olympics" Report, compiled by Planet Ark for National Recycling Week 2004, found at http://www.planetark.com/nrw/04RecyclingReport.pdf.

p. 54, e-waste is becoming a burden: see Harri Kalimo, *E-Cycling: Linking Trade and Environmental Law in the EC and the U.S.* (Ardsley, NY: Transnational Publishers, 2006).

p. 54, 300 million computers: Silicon Valley Toxics Association, "Poison PCs and Toxic TVs: E-waste Tsunami to Roll Across the US: Are We Prepared?," http://www.svtc.org/cleancc/pubs/ppcttv2004execsum.htm. See also Kevin Carmody, "U.S. Computer Makers Score Poorly on Toxic-Materials Report," *Austin American-Statesman*, 29 November 2001.

p. 55, "extended producer responsibility": For more on this regulatory trend see Margaret Walls, "Extended Producer Responsibility and Product Design," Working Paper, Resources for the Future (Washington, D.C., March 2006).

p. 55, Nokia . . . ahead of regulations: Tapio Takalo, Kirsi Sormunen, Harri Kalimo (Nokia), interview by author, Espoo, Finland, 3 September 2004.

p. 55, *Fortune 100* list: L. Michael Cacace et al., "The Hot 100," *Fortune*, 5 September 2005.

p. 56, 150 million cases of skin cancer: U.S. EPA, "Regulatory Impact Analysis: Protection of Stratospheric Ozone," 1987.

p. 57, fisheries are overexploited: "Global Oceans Conference Finds Progress Slow," *Environmental News Service*, 31 January 2006.

p. 57, About 20 percent of the world's coral reefs: Australian Institute of Marine Science, *Status of Coral Reefs of the World: 2004: Volume 1*, http://www.reefbase.org/References/ref_literature_detail.asp?refID=23038. In the Caribbean, 80 percent of the coral reef coverage has been lost over the last 30 years: see Lauren Morello, "Deaths of Corals, Sea Lions Signal Growing Marine Crisis," *Greenwire*, 17 February 2006.

p. 57, mouth of the Mississippi River: Richard Manning, "The Oil We Eat," *Harper's Magazine*, 308 (2004): 3.

p. 57, North Atlantic Fishery Collapse: See "Heavy Seas," *Economist.com*, 30

December 2003; National Center for Policy Analysis webpage, www.DebateCentral.org; http://www.un.org/events/tenstories/story.asp?storyID=800.

p. 58, area equal to Texas, California, and New York: Calculation by authors based on data from Food and Agriculture Organization of the United Nations, Forestry Department, "Global Forest Resources Assessment 2005: 15 Key Findings," 2006.

p. 60, deserts are growing . . . likely causes: Sachiko Kuwabara-Yamamoto, "The 'Rio' Environmental Treaties Colloquium," *Pace Environmental Law Review*, 13 (1995): 111.

p. 60, a dust cloud blew in: Howard French, "China's Growing Deserts are Suffocating Korea," *New York Times*, 14 April 2002; Soh Ji-young, "Worst Ever Yellow Dust Expected this Spring," *Korea Times*, 20 February 2004.

CHAPTER 3. WHO'S BEHIND THE GREEN WAVE?

p. 66, understand the who: For more on stakeholders, see R. K. Mitchell, B. R. Agle, and D. J. Wood, "Toward a Theory of Stakeholder Identification and Salience: Defining the Principle of Who and What Really Counts," *Academy of Management Review*, 22:4 (1997): 853–886.

p. 68, privately run initiatives that take on the feel of regulations: see Benjamin Cashore, *Governing Through Markets: Forest Certification and the Emergence of Non-state Authority* (New Haven: Yale University Press, 2004).

p. 69, 6 percent trusted business: Richard Edelman, "Rebuilding Public Trust Through Accountability and Responsibility," Address, Ethical Corporation Magazine Conference, New York, New York, 3 October 2002.

p. 69, "Environmentalists have smartened up their act . . .": Leyla Boulton, "Greens Follow Suit," *The Financial Times*, 30 December 1997.

p. 70, stood outside Fenway Park: Mark Buckley, "Successful Collaboration in Action," Presentation, Ethical Corporation NGO and Business Partnership Conference, Washington, D.C., 25 May 2005.

p. 70, Dell's wife faced angry protestors: Pat Nathan, Bryant Hilton (Dell), interview by author, Round Rock, TX, 16 October 2004.

p. 70, Vail . . . was the victim of arson: "Group Claiming Credit for Vail Fires Says the Aim Was to Help Lynx," CNN.com, 22 October 1998.

p. 70, Aspen Skiing Company recently committed: Paul Tolme, "Blowing in the Wind," *Newsweek*, 13 March 2006.

p. 70, watchdog groups rate: For ratings of resorts, see www.skiareacitizens.com.

p. 72, Kyoto Protocol: For the latest count of signatories, see http://unfccc.int/files/essential_background/kyoto_protocol/application/pdf/kpstats.pdf.

p. 72, Seattle mayor Greg Nickels: Kathy Mulady, "Seattle Dreams of 'Green' Team: Mayor Urging Other U.S. Cities to Enact Kyoto Protocol," *Seattle Post-Intelligencer Reporter*, February 2005.

p. 73, "from regulating outputs to regulating inputs": Wayne Balta, "Supply Chain Strategies in the Information Technology Industry," Speech, World Environment Center Conference "Clean Production Strategies in the Supply Chain," Toronto, Canada, 10 March 2006.

p. 75, compliance costs for cutting acid rain: Robert Percival et al., *Environmental Regulation: Law, Science, and Policy, 4th edition* (New York: Aspen Publishers, 2003), 544.

p. 75, price per ton: For a price chart, see Percival et al., *Environmental Regulation*, p. 545. In the last few years the price of a ton of sulfur dioxide has risen dramatically to over $600. One cause is the rising price of low-sulfur coal. See Tom Fowler, "Rising Price of Pollution: Cost of Buying Sulfur Dioxide Allowances Rises Sharply," *Houston Chronicle*, 28 March 2005, posted at www.chron.com.

p. 75, "appalling track record": Nelson Schwartz, "Inside the Head of BP," *Fortune*, 26 July 2004.

p. 76–77, disclose much more about environmental risks: See Ann Johnston and Angeles T. Rodriguez, "Environmental Disclosure: Come Clean in the Green Wave or Face the Heat," *Natural Resources & Environment*, Winter 2006, 3–8.

p. 77, include environmental liabilities: Gregory A. Bibler and Christopher P. Davis, "Disclosing Environmental Liabilities in the Wake of Sarbanes-Oxley," *Metropolitan Corporate Counsel*, April 2003.

p. 79, plans for a new coal plant: "Australia Planned Coal Station Cancelled Over Greenhouse Concerns," Ethicalcorp.com, 17 November 2005.

p. 83, obscure the emerging science on global warming: Timothy Gardner, "Global Climate Coalition Battles Kyoto Treaty," *Reuters*, 8 November 2000.

p. 84, Star-Kist's pledge: In the end, Star-Kist's advantage over the competition was very short-lived. For an in-depth analysis of this case study, see Forest Reinhardt, *Down to Earth* (Boston: Harvard Business School Press, 2000), 31–44.

p. 84, Wal-Mart . . . has suppliers across the world scrambling: Melanie Warner, "Wal-Mart Eyes Organic Foods," *New York Times*, 12 May 2006, A1.

p. 84, "Greening the supply chain": See United Nations Industrial Development Organization (Eds.), *Sustainable Supply Chains: The Global Compact Case Studies Series*, 2005, http://www.unido.org/doc/42222.

p. 85, Boise Cascade: "Boise Cascade to End Its Purchase of Wood Products from Endangered Forests," *Wall Street Journal*, 15 September 2003.

p. 85, public campaign against its Victoria's Secret line: Jeremy Caplan, "Paper War: Environmentalists Take on Victoria's Secret for Mailing More Than 1 Million Catalogs a Day," *Time*, 12 December 2005.

p. 86, CEO peers: The days of golf club connections are hardly over. See Landon Thomas, "A Path to a Seat on the Board? Try the Fairway," *New York Times*, 11 March 2006, A1.

p. 87, **Pressler's daughter asked:** Paul Pressler, "Keynote II," Speech, Business for Social Responsibility Conference, New York, New York, 10 November 2004.

p. 88, **A poll of U.S. evangelicals:** Paul Nussbaum, "Increasingly, Evangelists Are Embracing Environmentalism," *Knight Ridder Newspapers*, 25 May 2005.

p. 88, **a group of evangelical ministers . . . ran full-page ads:** See www.christiansandclimate.org for more information on the campaign.

p. 88, ***Ranger Rick* had started a postcard campaign:** Dave McLaughlin (Chiquita), conversation with author, 6 July 2004.

p. 89, **half a tunnel:** The story of Alcan's water diversion tunnel is based on Dan Gagnier (Alcan), interview with author, Vancouver, Canada, 31 March 2004; Simon Laddychuk, Paola Kistler (Alcan), conversation with author, 23 June 2004; and "Alcan Suing BC Government Over Cancelled Power Project (Kemano Completion Project on the Nechako River)," *Canadian Press Newswire*, 22 January 1997.

p. 90, **what employees need from a workplace:** The classic theory of evolving human needs is "Maslow's Hierarchy." For more on the role of this theory in sustainability, see Pieter Winsemius and Ulrich Guntram, *A Thousand Shades of Green* (London: Earthscan, 2002), 4–11.

p. 90, **business students would give up $13,700 per year:** David B. Montgomery and Catherine A. Ramus, Research Paper No. 1805, "Corporate Social Responsibility Reputation Effects on MBA Job Choice," May 2003.

p. 91, **"Good people won't want to work for us":** Steve Ramsey (GE), interview by author, 13 January 2005.

p. 91, **CEO Anne Mulcahy:** Anne Mulcahy, "Keynote IV," Speech, Business for Social Responsibility Conference, New York, New York, 11 November 2004.

p. 91, **dollars tell only part of the story:** For more on how the investment community affects sustainability, see S. Schmidheiny, F. J. Zorraquin, and the World Business Council for Sustainable Development, *Financing Change: The Financial Community, Eco-Efficiency, and Sustainable Development* (Cambridge, MA: MIT Press, 1996).

p. 92, **indicator of good general management:** Although there is much debate about whether SRI investments do better or worse than the market at large, Innovest, a New York–based entity that ranks corporate sustainability, makes a compelling case that companies that manage environmental issues well manage all issues better—they're just well-run companies. Environmental excellence becomes another indicator investors can use to spot likely superior performers.

p. 92, **Merrill Lynch report picked auto stocks:** "Merrill Lynch, World Resources Institute Analyze Climate Change Investment Opportunities," SocialFunds.com, http://www.greenbiz.com/news/news_third.cfm?NewsID=28304.

p. 92, "people will take the FTSE4Good seriously": Sarah Ryle, "That's Not Sir to You," *The Guardian*, 24 February 2002.

p. 92, state treasurers and comptrollers: "Investors at UN Talk About Climate Change," *Associated Press*, 11 May 2005.

p. 92–93, Two of the three largest pension funds: Alex Kaplun, "Calif. Treasurer to Call for Investment in Enviro-friendly Companies," *Greenwire*, 3 February 2004.

p. 93, Carbon Disclosure Project: See http://www.cdproject.net.

p. 94, Mindy Lubber, executive director of Ceres: Christopher Rowland, "Greening of the Boardroom: Socially Conscious Investors Get Results on Global Warming," *Boston Globe* online, 31 March 2005, www.boston.com.news.

p. 94, "The real issue for insurers is natural disasters": Dean Calbreath, "Changes in Climate Pose Greatest Challenge for Insurers, Say Experts from Around World," *San Diego Union Tribune*, 23 April 2004.

p. 94, 2003 heat wave: James Reynolds, "Urgent Action Is Called for Now, Says Blair in Dire Global Warning," *The Scotsman*, 15 September 2004.

p. 94, worldwide economic losses from natural disasters: 2004 and 2005 expenses for natural disasters from "Natural Disasters Made 2005 Costliest for Insurance Industry: Munich Re," *Agence France-Presse*, 29 December 2005.

p. 95, Equator Principles: See Benjamin C. Esty, "The Equator Principles: An Industry Approach to Managing Environmental and Social Risks," Harvard Business School Case Study N9-205-11, 2005.

p. 96, "tip of the iceberg": André Abadie, interview by author, New York, NY, 2 June 2005.

p. 96, JPMorgan . . . told the world: Jim Carleton, "J.P. Morgan Adopts 'Green' Lending Policies," *Wall Street Journal*, 25 April 2005, B1.

p. 96, Citigroup . . . screen projects: Associated Press, "Citigroup Adopts Environmental Policy," www.nytimes.com, 22 January 2004.

PART TWO. STRATEGIES FOR BUILDING ECO-ADVANTAGE

CHAPTER 4. MANAGING THE DOWNSIDE

p. 105, this strategy has saved DuPont $2 *billion:* Paul Tebo, Dawn Rittenhouse, and Ed Mongan (DuPont), interview by author, Wilmington, DE, 5 March 2004.

p. 106, "wet processing": John Caffall, Terry Maloney, and Julia Bussey (AMD), conversation with author, 27 May 2004, and correspondence with author, 27 February 2006.

p. 106, Timberland redesigned its shoe boxes: Terry Kellogg (Timberland), conversations with author, 3 December 2003, 25 March 2005, and 8 April 2005.

p. 106, IBM . . . saving $115 million: "IBM Cuts CO_2 Emissions by More Than 1 Million Tons, Saving $115 Million," *Greenbiz.com*, 30 September 2005.

p. 106, computerized sprinkler system: Jo Ann Steinmetz, "SJ Headquarters Use of Environmental Tech Also Cut Costs," *San Jose Mercury News*, 17 January 2006.

p. 106–107, Pollution Prevention Pays: History of 3P program and discussion in "What's Wrong with Abatement" comes from Kathy Reed, Keith Miller (3M), interview by author, Minneapolis, MN, 24 May 2004; and Thomas Zosel, "Pollution Prevention," *Vital Speeches of the Day*, 65:8 (1 February 1999): 243.

p. 106–107, "Anything not in a product is considered a cost": Kathy Reed (3M), interview by author, Minneapolis, MN, 24 May 2004. Also Zosel, "Pollution Prevention."

p. 107, savings of 2.2 billion pounds of pollutants: See http://solutions.3M.com for 3M's estimates on total reductions.

p. 107, "new ways to reduce natural gas use": Jim Omland (3M), interview by author, Minneapolis, MN, 24 May 2004.

p. 108, Rhone-Poulenc . . . found a market for . . . a by-product: Michael Porter and Claas Van der Linde, "Green and Competitive: Ending the Stalemate," *Harvard Business Review*, 73 (1995): 120–134.

p. 108, spirit of "industrial ecology": The field of industrial ecology is discussed at length in T. E. Graedel and B. R. Allenby, *Industrial Ecology*, 2nd ed. (Upper Saddle River, NJ: Prentice-Hall, 2003). See also R. Lifset. and T. E. Graedel, "Industrial Ecology: Goals and Definitions," in *Handbook of Industrial Ecology*, edited by R. Ayres and L. Ayres (Cheltenham, UK: Edward Elgar, 2002); and D. Allen, "Using Wastes as Raw Materials: Opportunities to Create an Industrial Ecology," *Hazardous Waste and Hazardous Materials*, 10, no. 3: 273–277.

p. 109, waste exchange websites: For more on exchanging by-products, see EPA document 530K94003, "Review of Industrial Waste Exchanges," National Environmental Publications Information System, 1994, http://nepis.epa.gov/pubtitle.htm; see also Daniel C. Esty, "Environmental Protection in the Information Age," *NYU Law Review*, 79:1 (2004): 115–211.

p. 109, Dow Chemical set employees' computers to shut down . . . Staples saved $6 million: Nicholas Varchaver, "How to Kick the Oil Habit," *Fortune*, 23 August 2004.

p. 109, Kinko's retrofitted over 95 percent: Larry Rogero (Kinko's), conversations with author, 25 November 2003 and 12 December 2003.

p. 110, scrapped the program: Albin Kaelin (Rohner), interview by author, Heerbrugg, Switzerland, 25 August 2004.

p. 110, "kept us competitive": Reed, interview.

p. 110, "We wouldn't have made it": Ray Anderson (Interface), interview by author, New Haven, CT, 6 March 2005.

p. 111, one of the world's largest polluters: The story of DuPont's reaction to the Toxic Release Inventory and waste reduction efforts from Paul Tebo, et al., interview.

p. 111, up to 20 percent: Dan Gagnier (Alcan), interview by author, Vancouver, Canada, 31 March 2004; also correspondence with author, 25 February 2005.

p. 112, saved the company over $1 million: Paul Murray (Herman Miller), interview by author, 4 May 2004.

p. 112, chemicals that are known dangers: For more details on the health effects of toluene and other substances, see http://www.atsdr.cdc.gov/HEC/CSEM/toluene/physiologic_effects.html.

p. 112, water-based adhesives: Terry Kellogg (Timberland), conversations with author, 3 December 2003, 25 March 2005, and 8 April 2005; and Betsy Blaisdell (Timberland), conversation with author, 6 March 2006.

p. 113, save the auto companies $20 million: "DuPont Technology to Receive U.S. EPA's Clean Air Excellence Award; Innovative SuperSolids(TM) Coatings Technology Reduces Emissions by More Than 25 Percent for Automotive Finishes," *PR Newswire*, 6 March 2003.

p. 113, "flat packaging": Thomas Ivarsson (IKEA), interview by author, Almhult, Sweden, 24 August 2005. Additional information on the KLIPPAN sofa in IKEA's report, "Social and Environmental Responsibility," 2005, p. 27.

p. 113, Dell has upped its average truck load: Dick Hunter (Dell), interview by author, Parmer, TX, 16 December 2004.

p. 113, save $110,000: See http://solutions.3M.com for more details.

p. 114, Spidey Signals toy: Jim Fitzgerald, "Kellogg to Drop Mercury-battery Toys after Spider-Man Promotion," *Associated Press*, as reported on http://www.maineenvironment.org/toxics/spidey_APstory.htm, 16 July 2004.

p. 114, A few years before Kellogg's: Information on McDonald's Anticipatory Issues Management and their move away from mercury batteries from Bob Langert (McDonald's), interview by author, Oak Brook, IL, 28 July 2004. For some context on scale, note that McDonald's spends twice as much on toys as it does on soda (from draft version of the Supplemental Report to McDonald's 2004 Corporate Responsibility Report).

p. 115, Oprah interviewed a vegetarian: "Oprah Accused of Whipping Up Anti-Beef Lynch Mob," CNN.com, 21 January 1998, http:///www.cnn.com/US/9801/21/oprah.beef/.

p. 116, Shell uses scenario planning: "The Shell Global Scenarios to 2025," http://www.shell.com/static/royal-en/downloads/scenarios/exsum_23052005.pdf.

p. 116, experts at the Institute of Risk Management: IRM, AIRMIC, and ALARM, "A Risk Management Standard," Institute of Risk Management, 2002.

p. 118, McDonald's Hungary: Else Krueck (McDonald's), conversation with author, 28 July 2004.

p. 118, internal process called Greenlist: David Long, "Green List," Presentation, Business for Social Responsibility Annual Conference, Washington, D.C., 3 November 2005; and correspondence with author, 7 March 2006.

p. 118, Nokia has reviewed 30,118 components: Harri Kalimo (Nokia), correspondence with author, 27 February 2006.

p. 119, "We are in a carbon-constrained world": Jeff Immelt, launch event for ecomagination, Washington, D.C., 9 May 2005.

p. 119, Electrolux announced a partnership with Toshiba: "Toshiba, Electrolux to Form Tie-Up," *Daily Yomiuri*, 25 May 1999.

p. 120, Japan's "Top Runner" product labeling program: See Japan's "Top Runner" labeling program at http://www.eccj.or.jp/top_runner/index_contents_e.html.

p. 120, "a seat at the table": Darcy Frey, "How Green Is BP?" *New York Times*, 8 December 2002.

p. 120, Nokia found it very useful: Kalimo, correspondence.

p. 121, Champion Paper thrived: Forest Reinhardt, "Champion International Corp.: Timber, Trade, and the Northern Spotted Owl," Harvard Business School Case 792017, 1991.

p. 121, DuPont initially fought the phaseout: The DuPont CFC story has been told numerous times, but perhaps most clearly by Harvard Business School Professor Forest Reinhardt in two works in particular: Richard H. K. Vietor and Forest Reinhardt, "Du Pont Freon Products Division," Harvard Business School Case 389111, 1989; and Forest Reinhardt, *Down to Earth* (Boston: Harvard Business School Press, 2000), 61–65.

CHAPTER 5. BUILDING THE UPSIDE

p. 122, "I never imagined I'd be talking about the environment": Jeff Immelt, launch event for ecomagination, Washington, D.C., 9 May 2005.

p. 122, Early results are very promising: Fiona Harvey, "General Electric Bolsters 'Green' Revenues," *Financial Times*, 17 May 2006.

p. 124, IdleAire's efficiency innovation: Laurent Belsie, "Companies That Cut Pollution for Profit," *Christian Science Monitor*, 22 April 2002, posted at http://www.csmonitor.com/2002/0422/p16s01-wmcr.html.

p. 124, trash can called BigBelly: Karen Spaeder, "Turn Environmental Problems Into Opportunities," entrepreneur.com, reprinted on www.msnbc.com, 17 March 2006.

p. 124, Scott McNealy said: Matt Stansberry, "Server Specs: Dept. of 'Green' Computing," *SearchDataCenter.com*, 3 October 2005.

p. 124, Dell's Asset Recovery System: Dell is not the only (or even the first) PC or electronics manufacturer doing some form of "take-back." IBM, for example, has been working with customers on end-of-life issues since as early as 1991.

p. 125, Dell refurbishes and reuses some parts: Tod Arbogast (Dell), conversation with author, 30 March 2005.

p. 125, a new technique for recycling polyester: Paul Tebo, Dawn Rittenhouse, and Ed Mongan (DuPont), interview by author, Wilmington, DE, 5 March 2004; and conversation with author, 3 June 2004.

p. 126, VOC-free tapes often melted: Kathy Reed, Keith Miller (3M), interview by author, Minneapolis, MN, 24 May 2004.

p. 126, reusable mugs: Bob Langert (McDonald's), interview by author, Oak Brook, IL, 28 July 2004.

p. 126, When a nurse asked: Kathy Reed, Keith Miller, Jim Omland (3M), interview by author, Minneapolis, MN, 24 May 2004.

p. 127, marketing a new, cleaner-burning gasoline: Mark Weintraub (Shell), interview by author, The Hague, Netherlands, 20 October 2004.

p. 128, "another door to sales": Reed et al., interview.

p. 128, GrupoNueva solidifies its role: Maria Emilia Correa, conversation with author, 1 June 2004.

p. 129, "second or third button": Weintraub, interview.

p. 129, $4-million bump in sales: Antron was owned by DuPont and is now part of Koch Industries. Sales increase from "2003 DuPont Sustainable Growth Excellence Awards," internal document provided by Dawn Rittenhouse (DuPont).

p. 130, Timberland . . . nutritional content label: "Timberland Introduces New Packaging Initiative," January 27, 2006, http://greenbiz.com/news/news_third.cfm?NewsID=30215.

p. 130, certification and labeling: James Salzman, "Informing the Green Consumer: The Debate Over the Use and Abuse of Environmental Labels," *Journal of Industrial Ecology* 1:2 (1997): 11–21.

p. 130, established by private entities: See MSC website (www.msc.org) and FSC website (www.fscus.org).

p. 132, "We forbade the sales force": Ray Anderson (Interface), interview by author, New Haven, CT, 6 March 2005.

p. 133, Toyota set out to design the "21st century car": We owe much of the behind-the-scenes detail on the development of the Toyota Prius to Jeffrey Liker's *The Toyota Way* (New York: McGraw-Hill, 2004).

p. 133, value innovation: The concept of value innovation is credited to INSEAD professors Chan Kim and Renée Mauborgne; see "Value Innovation: The Strategic Logic of High Growth," *Harvard Business Review*, January-February 1997, 103–112.

p. 133, made the competition irrelevant: See Chan Kim and Renée Mauborgne, *Blue Ocean Strategy: How to Create Uncontested Market Space and Make the Competition Irrelevant* (Boston: Harvard Business School Press, 2005).

p. 134, cooling as a service: See Paul Hawken, Amory Lovins, and L. Hunter Lovins, *Natural Capitalism: Creating the Next Industrial Revolution* (Boston: Back Bay Books, 1999).

p. 135, things didn't work out as planned: Anderson, interview.

p. 136, "Beyond Preposterous": Kenny Bruno, "BP: Beyond Petroleum or Beyond Preposterous?" *CorpWatch*, 14 December 2136, http://www.corp watch.org/article.php?id=219.

p. 136, "helios design and overall positioning": Chris Mottershead (BP), interview by author, London, England, 21 October 2004.

p. 136–137, "what you stand for": Mottershead, interview.

p. 136–137, dialed back its rhetoric: Darcy Frey, "How Green is BP?," *New York Times*, 8 December 2002.

p. 137, list of most admired CEOs: See "Britain's Most Admired Companies 2005," *Management Today*, 2 December 2005, 41; and "Most Admired League Table 2004: The Measures of Success," *Management Today*, 2 December 2004.

p. 137, gained over $3 billion in brand value: *Fortune* asked Landor, the brand consultancy that has for years valued brands with its BrandAsset Valuator, to provide it with a selection of brands that had increased in strength. Another firm, BrandEconomics, put a value on the economic value gained. The full analysis is in Al Ehrbar, "Breakaway Brands," *Fortune*, 31 October 2005, 153–170.

p. 137, "We don't have the recruitment problems": Mottershead, interview.

p. 137, Hefty's highly touted biodegradable garbage bag: Michael Jay Polonsky and Philip J. Rosenberger III, "Reevaluating Green Marketing: A Strategic Approach," *Business Horizons*, 44 (2001): 21–31.

p. 138, "Jeff picked the toughest": Lorraine Bolsinger, "The Story Behind GE's Ecomagination," Speech, Business for Social Responsibility Annual Conference, Washington, D.C., 4 November 2005.

p. 138, ecomagination product . . . GEnx jet engine: Information on the criteria for ecomagination products from Bolsinger, Speech, as well as Joe Malcoun (GreenOrder), conversation with author, 31 October 2005; Andrew Shapiro (GreenOrder), conversation with author, 5 March 2006; also correspondence with Peter O'Toole (GE), 28 February 2005.

p. 138, supermodels in a coal mine: To see all the ecomagination ads, visit http://www.ge.com/en/company/companyinfo/advertising/eco_ads.htm.

p. 139, Unilever's "Vitality" Positioning: Jeroen Bordewijk (Unilever), interview by author, Rotterdam, Netherlands, 19 October 2004; Chris Pomfret (Unilever), conversations with author, 19 October 2004 and 1 November 2004; Dierk Peters (Unilever), conversation with author, 2 November 2004.

p. 139–140, Levi's was buying 2 percent of its cotton from organic farmers: Joel Makower, "Nike Things Considered," http://makower.typepad.com/joel_makower/2005/03/nike_things_con.html.

p. 141, free of animal testing: Joan Harrison, "Acquiring the Socially Conscious Company," *Mergers and Acquisitions*, June 2000.

p. 141, "hell to pay": Richard Wells, correspondence with author, 23 February 2006.

PART THREE. WHAT WAVERIDERS DO

p. 144, **corporate DNA:** See opening letter from Jeffrey Immelt and Steve Ramsey (Vice President, Environmental Programs) in GE, "Being Responsible: GE 2004 Environmental, Health, and Safety Report," p. 3. http://www.ge.com.

CHAPTER 6. THE ECO-ADVANTAGE MINDSET

p. 145–146, **Fosbury Flop:** Mark Beech, "The Originals," *Sports Illustrated* 93, no. 5 (2000): 94–95.

p. 148, **the "strategic term":** Rick Paulson (Intel), interview by author, Chandler, AZ, 28 January 2004.

p. 148, **futures the team famously imagined:** The history of the Shell Scenarios group is described in Peter Schwartz, *The Art of the Long View* (New York: Doubleday, 1991).

p. 149, **The Grandchildren Test . . . :** Kate Sosnowchik and Joseph Fiksel, "Awakening a Sustainability Giant," *Green@Work* (Winter 2004), 14.

p. 150–151, **new and fast-growing market for Post-its:** Kathy Reed, Keith Miller, Gregory Anderson, Valerie Young (3M), interview by author, Minneapolis, MN, 24 May 2004.

p. 152, **Green Cost Accounting:** See Marc Epstein, *Measuring Corporate Environmental Performance: Best Practices for Costing and Managing an Effective Environmental Strategy* (New York: McGraw-Hill, 1995).

p. 152, **"solvent-less technology":** Reed et al., interview.

p. 153, **"most sophisticated biodiversity strategy out there":** Glenn Prickett (Conservation International), conversation with author, 25 October 2004.

p. 155, **how his team creates value:** Gudmond Vollbrecht (IKEA), interview by author, Gelterkindern, Switzerland, 27 August 2004.

p. 156, **Some 11 million cartridges:** "HP Expands Planet Partners Program," *Green@Work* (May/June 2004), 10; see also www.hp.com.

p. 157, **eliminating the use of lead-based stabilizers:** Maria Emilia Correa (GrupoNueva), conversations with author, 1 June 2004 and 3 June 2004. For more on GrupoNueva, see also Monica Araya, "To Tell or Not to Tell? Determinants of Environmental Disclosure and Reporting in Corporate Latin America," (Ph.D. Dissertation, Yale University, 2006), 275–284.

p. 158, **Apollo 13's April 1970 lunar mission:** See Gene Kranz, *Failure Is Not an Option: Mission Control from Mercury to Apollo 13 and Beyond* (New York: Simon & Schuster, 2000). For basic details on the Apollo 13 story, see http://en.wikipedia.org/wiki/Apollo_13#Problem and Jim Lovell's account on http://history.nasa.gov/SP-350/ch-13-1.html.

p. 158–159, **Ed Woolard, sent a clear message . . . nylon intermediates:** Paul Tebo, Dawn Rittenhouse, and Ed Mongan (DuPont), interview by author, Wilmington, DE, 5 March 2004.

p. 159, 10 million pounds of toxics: Toxics Release Inventory data available at www.rtk.net.

p. 159, tough on everyone, including suppliers: Pat Nathan (Dell), interview by author, Round Rock, TX, 16 December 2004; Terry Kellogg (Timberland), conversations with author, 3 December 2003, 25 March 2005, and 8 April 2005; Larry Rogero (Kinko's), conversations with author, 25 November 2003 and 12 December 2003.

p. 160, Troon North Golf Club: "The *Golf Magazine* Golf Course Guide," *Golf Magazine*, GolfCourse.com, http://www.golfcourse.com/search/coursedtl_ga.cfm?source=GA&courseid=118.

p. 160, water needed for just one Arizona golf course: Calculations based on golf course water use figures from "War Over Water," *Golf Course News*, 20 January 2005, and roster of Arizona courses on www.1golf.com/az/index.htm. For data on water needs globally, see http://www.infoforhealth.org/pr/m14/m14chap2_2.shtml, and A. Marcoux, *Population and Water Resources* (Rome: United Nations Food and Agriculture Organization Sept. 1994), 4–33.

p. 160, "hottest issue here is water use": Paulson, interview.

p. 161, "it's what the *perception* is": Paulson, interview.

p. 162, Shapiro put sustainability at the very center: Joan Magretta, "Growth Through Sustainability (an Interview with Monsanto's CEO, Robert Shapiro)," *Harvard Business Review* 75:1 (1997): 78–88.

p. 163, little wooden pencils: Bob Kay (IKEA), interview by author, Paramus, NJ, 20 January 2005.

p. 164, Herman Miller's founder: Herman Miller, "Journey Toward Sustainability," http://www.hermanmiller.com/CDA/SSA/Category/0,1564,a10-c605,00.html.

p. 164, "It's the right thing to do": Carly Fiorina, "Opening Keynote," Speech, Business for Social Responsibility Conference, Los Angeles, CA, 12 November 2003.

p. 164, Brian Walker writes: Herman Miller, "The Environment: A Better World Together," http://www.hermanmiller.com/CDA/SSA/Category/0,1564,a10-c382,00.html.

CHAPTER 7. ECO-TRACKING

p. 168, total footprint: For more on the concept of environmental footprint, see Mathis Wackernagel and William Rees, *Our Ecological Footprint: Reducing Human Impact on the Earth* (Gabriola Island, BC: New Society Publishers, 1996).

p. 169, WaveRiders take the measure of their footprint seriously: Interestingly, BP is working to popularize this term through a series of ads that discuss the footprint concept.

p. 171, service business like the bank branch: If the focus of the LCA were on the bank itself, not the product, then a quite different analysis would be required. It would center not on the issues arising from its branches, but rather on the downstream emissions of those it finances. This fact drives the commitment of major banks to the Equator Principles and to looking at the environmental effects of the loans they make.

p. 173, spin out of control: LCA software is now available that can be a great help; see our website, www.eco-advantage.com, for more information. Each LCA is, however, a custom affair.

p. 173, "system conditions": The Natural Step has developed highly specific language for its system conditions, which reads as follows: "In the sustainable society, nature is not subject to systematically increasing . . . (1) . . . concentrations of substances extracted from the Earth's crust (e.g. cadmium and fossil CO_2), (2) . . . concentrations of substances produced by society (e.g. CFC's and endocrine disrupters), (3) . . . degradation by physical means (e.g. deforestation or overfishing); and people are not subject to conditions that systematically (4) undermine their capacity to meet their needs (e.g. from abuse of political and economic power)." For more on the Natural Step, see Brian Nattrass and Mary Altomare, *The Natural Step for Business* (Gabriola Island, BC: New Society Publishers, 1999).

p. 173, "backcasting from principles": Karl-Henrik Robert, conversation with author, 16 November 2004.

p. 173–174, manage any aspect of business performance: For more information on performance measurement, see Marc J. Epstein, J.F. Manzoni, eds., *Performance Measurement and Management Control: Superior Organizational Performance* (New York: Elsevier Science Ltd., 2004); see also Daniel C. Esty, "Why Measurement Matters," in *Environmental Performance Measurement: The Global 2001–2002 Report* (Daniel C. Esty and Peter Cornelius, eds.) New York: Oxford University Press (2002).

p. 174, "balanced scorecard": Balanced scorecard is a strategic management tool developed by Harvard Business School's Robert S. Kaplan and David P. Norton; see their *The Balanced Scorecard: Translating Strategy into Action* (Cambridge: HBS Press, 1996).

p. 174, Coca-Cola tracks: "The World's Biggest Drinks Firm Tries to Fend Off Its Green Critics," Economist.com, 6 October 2005.

p. 176, Herman Miller built a materials database: Gabe Wing and Scott Charon (Herman Miller), interviewed by author, Zeeland, MI, 3 May 2004. The scoring system they used was based on analysis by McDonough/Braungart (MDBC).

p. 177, "we can answer any questions with data": Don Brown (Dell), interview by author, Parmer, TX, 16 December 2004.

p. 177, Shareholder Value Added: SVA is virtually identical to another commonly used metric, Economic Value Added. It's defined as "a value-based performance measure of a company's worth to shareholders. The basic calculation is net operating profit after tax (NOPAT) minus the cost of capital from the issuance of debt and equity, based on the company's weighted average cost of capital (WACC)" (from http://www.investopedia.com/terms/s/shareholdervalueadded.asp.)

p. 177–178, SVA/pound against the Price/Earnings ratio: Paul Tebo, Dawn Rittenhouse, and Ed Mongan (DuPont), interview by author, Wilmington, DE, 5 March 2004.

p. 178, ranking of countries: "2006 Environmental Performance Index," Yale Center for Environmental Law and Policy (2006), available at www.yale.edu/epi.

p. 180, ISO 14000: See Aseem Prakash, "A New-Institutionalist Perspective on ISO14000 and Responsible Care," *Business Strategy and the Environment* 8 (1999): 322–335.

p. 180, Help in implementing ISO 14000: See www.wbcsd.org and www.gemi.org.

p. 180, European Union's Eco-Management and Audit Scheme: See www.europa.eu.int/comm/environment/emas/index_en.htm.

p. 181, total tab over $12 billion: "Exxon back in court over 1989 Valdez Spill Fine," MSNBC News service, 27 January 2006, see http://msnbc.msn.com/id/11059801/%23storycontinued; ExxonMobil has spent $3 billion to settle state and federal charges. It continues to appeal the judgment against it in favor of fishermen, which accrued interest now approaching $9 billion.

p. 181, corporate crisis management team: Wayne Balta, "Supply Chain Strategies in the Information Technology Industry," Speech, World Environment Center Conference: "Clean Production Strategies in the Supply Chain," Toronto, Canada, 10 March 2006.

p. 182, Deal Review . . . on-site day care: Colleen Conner, Jack Campbell, Mark Stoler, Stephen Ramsey (GE), interview by author, Fairfield, CT, 13 January 2005.

p. 182, Whole books have been written: For example, see Jem Bendell, ed., *Terms for Endearment*, (Sheffield, UK: Greenleaf Publishing, 2000). See eco-advantage.com for more information and examples of the types of partnerships we cover here.

p. 183–185, unlikely partnership: The discussion of Chiquita and Rainforest Alliance partnership on banana production is based on Dave McLaughlin (Chiquita), conversations with author, 6 July 2004 and 23 July 2004; and Chris Wille (Rainforest Alliance), conversation with author, 3 August 2004. See also the very helpful book Gary Taylor and Patricia J. Scharlin, *Smart Alliance: How a Global Corporation and Environmental Activists Transformed a Tarnished Brand* (New Haven: Yale University Press, 2004), see Appendix C for sample environmental performance report.

p. 184, "strange bedfellows": Glenn Prickett, "Strange Bedfellows: Can Partnerships between Corporations and Non-Governmental Organizations Save the Environment?" Presentation at seminar at the New America Foundation, 20 November 2002.

p. 185, productivity is up 27 percent: Tensie Whelan (Rainforest Alliance), "Chiquita's Cost/Productivity: Are Corporate Responsibility Programs an Added Cost?" Presentation document, correspondence with author, 22 February 2006.

p. 185, **use of pesticides:** Organic bananas are impossible in Central America due to one fungus that requires chemical herbicides.

p. 185, **"danced with the devil":** Dave McLaughlin (Chiquita), conversation with author, 6 July 2004.

p. 185–186, **The other gold star:** Information on McDonald's partnerships comes from Bob Langert, Bruce Feinberg, Samantha Sturhahn (McDonald's), Kenneth Krause (Perseco), interview by author, Oak Brook, IL, 28 July 2004; Phylis Antonacci (OSI Group, a large beef supplier to McDonald's), conversation with author, 28 July 2004; and "McDonald's Socially Responsible Food Supply Initiative Progress Update," July 2004, internal document.

p. 187, **"They crawled all over our operations":** Dan Gagnier (Alcan), interview by author, Vancouver, Canada, 31 March 2004.

p. 188, **Project XL . . . a "huge win":** Tim Mohin, Rick Paulson, Dave Olney (Intel), interview by author, 28 January 2004; see also http://www.epa.gov/ProjectXL.

p. 188, **IPP focuses on the life-cycle impact:** Harri Kalimo (Nokia), correspondence with author, 27 February 2006; see also http://europa.eu.int/comm/environment/ipp/home.htm.

p. 189, **The Tar Sands:** Mark Weintraub (Shell), interviews by author, The Hague, Netherlands, 12 March 2004 and 20 October 2004; see also "Crude Realities: Mining the Future, the Dark Magic of Oil Sands," *Fortune*, 3 October 2005.

p. 192, **typology of NGOs:** John Elkington and Shelly Fennell, "Partners for Sustainability," in *Terms for Endearment*, ed. Jem Bendell (Sheffield, UK: Greenleaf Publishing, 2000).

CHAPTER 8. REDESIGNING YOUR WORLD

p. 195, **Eco-efficiency isn't good enough:** See in particular, William McDonough and Michael Braungart, *Cradle to Cradle* (New York: North Point Press, 2002).

p. 198, **a product they call Climatex:** The story of Climatex fabric stems from Albin Kaelin (Rohner), interview by author, Heerbrugg, Switzerland, 25 August 2004, and from "The Story of E," a document describing the fabric and its history, provided by Rohner.

p. 199, **Hitachi adopted a DfE strategy:** Daniel C. Esty and Michael E. Porter, "Industrial Ecology and Competitiveness: Strategic Implications for the Firm," *Journal of Industrial Ecology* 2:1 (1998): 38.

p. 199, **Dow Chemical:** Esty and Porter, "Industrial Ecology," 37.

p. 199–200, waste-to-energy plant: Kris Manos, Kent Gawart, Paul Murray (Herman Miller), interview by author, Zeeland, MI, 4 May 2004.

p. 200, Dutch flower industry developed: Esty and Porter, "Industrial Ecology," 38.

p. 200, "industrial symbiosis": See Marian R. Chertow, "Industrial Symbiosis: Literature and Taxonomy," *Annual Review of Energy and the Environment* 25 (2201): 313–337.

p. 201, LEED standards: More on the U.S. Green Building Council and its LEED certification program can be found at www.usgbc.org. New laws are mandating buildings meet LEED standards; see "USGBC, Engineering Groups Partner on Baseline Green Building Standard," *Greenbiz.com*, 16 February 2006.

p. 201, lowered utility costs: Len Pilon (Herman Milller), interview with author, Zeeland, MI, 3 May 2004.

p. 202, supply chain questions: For more on supply chains, see United Nations Industrial Development Organization (eds.), *Sustainable Supply Chains: The Global Compact Case Studies Series*, 2005, http://www.unido.org/doc/42222.

p. 202, suppliers that do poorly on their audits: Wayne Balta, "Supply Chain Strategies in the Information Technology Industry," Speech, World Environment Center Conference "Clean Production Strategies in the Supply Chain," Toronto, Canada, 10 March 2006; and correspondence with author, 13 March 2006.

p. 202–204, The IKEA Way: The details of IKEA's IWAY program come from Thomas Bergmark, Dan Brännström (IKEA), interview by author, Helsingborg, Sweden, 23 August 2004; Olle Blidholm (IKEA), interview by author, Copenhagen, Denmark, 23 August 2004; Gudmond Vollbrecht (IKEA), interview by author, Gelterkindern, Switzerland, 27 August 2004. Some details are also drawn from "IWAY Evaluation Checklist" edition 2.1, internal document provided by Thomas Bergmark.

p. 203, IKEA suppliers from Mexico: Olle Blidholm (IKEA), interview by author, 25 August 2004. See also IKEA's "Social and Environmental Responsibility Summary Report, 2004," p. 15.

CHAPTER 9. INSPIRING AN ECO-ADVANTAGE CULTURE

p. 206, In 1997, a Conoco oil tanker: Paul Tebo, Dawn Rittenhouse, and Ed Mongan (DuPont), interview by author, Wilmington, DE, 5 March 2004. See also "Crude-laden Double-hulled Tanker Survives Gash without a Spill," *Oil and Gas Journal*, 24 November 1997, 44.

p. 208, "cross the U.S. continent on one full tank of gas": Jathon Sapsford, "Toyota Adds to Executive Ranks as It Ramps up Research Efforts," *Wall Street Journal*, 27 June 2005.

p. 208, zero liquid effluent: "Environmental Report," Unilever, 2003.

p. 209, Michael Porter has argued: Michael Porter, "America's Green Strategy," *Scientific American* 264 (1991): 168. See also Michael Porter and Claas Van der Linde, "Green and Competitive: Ending the Stalemate," *Harvard Business Review* 73 (1995): 120–134.

p. 209, "Perfect Vision": Kris Manos (Herman Miller), interview by author, New Haven, CT, 30 November 2004.

p. 209, "it was time the company did much better": Dan Gagnier (Alcan), interview by author, Vancouver, Canada, 31 March 2004.

p. 210, "If you set goals": Tebo, interview.

p. 210, "Beyond Zero": Rick Lawrence (Alcan), conversation with author, 31 August 2004.

p. 211, "strategic thinking and transformation": Rittenhouse, interview.

p. 211, "Once you go public": Tebo, interview.

p. 212, slashes the hurdle rate: Kathy Reed and Keith Miller (3M), interview by author, Minneapolis, MN, 24 May 2004.

p. 212, ten- to fifteen-year payback: Thomas Bergmark, (IKEA), interview by author, Helsingborg, Sweden, 23 August 2004.

p. 212, issues down the road: Jim Omland (3M), interview by author, Minneapolis, MN, 24 May 2004.

p. 212–213, "I have the go-ahead": Bob Langert (McDonald's), interview by author, 28 July 2004.

p. 213, "no flaring" policy: Peter Davies, Chris Mottershead (BP), interview by author, London, England, 11 March 2004.

p. 213–214, redesign of its shoe boxes: Terry Kellogg (Timberland), conversations with author, 3 December 2003, 25 March 2005, and 8 April 2005.

p. 214, options for retrofitting: Kellogg, conversations.

p. 214, breaks even on the total recycling program: Paul Murray (Herman Miller), interview by author, 4 May 2004.

p. 215, internal trading system: Peter Davies, Chris Nicholson, and Chris Mottershead (BP), interview by author, London, England, 11 March 2004.

p. 215, "it's useful as a flag-raising exercise": Garth Edward, interview by author, London, England, 12 March 2004.

p. 215, "three things well": Chris Mottershead, interview by author, London, England, 11 March 2004.

p. 216, Rohner Textil put a tax on itself: Albin Kaelin (Rohner), interview by author, Heerbrugg, Switzerland, 25 August 2004.

p. 216, "green tags": Jerry Akers (Herman Miller), interview by author, Zeeland, MI, 4 May 2004.

p. 216, Hyperion Software: Erin White and Jeffrey Ball, "A Green Perk Offered for Green Car: Hyperion Gives Employees $5,000 to Purchase Vehicle," *Wall Street Journal*, 29 November 2004, B4.

p. 217–219, Herman Miller's Environmental Quality Action Team: All of the details on the EQAT are from Kris Manos, Kent Gawart, Paul Murray, and Brian Walker (Herman Miller), interview by author, Zeeland, MI, 4 May 2004.

p. 219–220, Executive Stewardship program: Shaye Hokinson, Reed Content, Rich Weigand, and Phil Trowbridge (AMD), conversation with author, 14 March 2004; Craig Garcia (AMD), interview by author, 11 May 2004.

p. 220, "keeper of the CSR flame": Kellogg, conversation.

p. 220, Tough Love: GE's Session E: Stephen Ramsey and Mark Stoler (GE), interview by author, Fairfield, CT, 13 January 2005.

p. 221–222, "get out of the 'greeny' corner": Nicole Schneider (IKEA), interview by author, Copenhagen, Denmark, 23 August 2004.

p. 222, brainstorming meetings: Brainstorming sessions mentioned in Jeffrey Liker, *The Toyota Way* (New York: McGraw-Hill, 2004); see also Charles O. Holliday, Jr., Stephan Schmidheiny, and Philip Watts, *Walking the Talk: The Business Case for Sustainable Development*, (Sheffield, UK: GreenLeaf, 2002).

p. 222, janitors together with engineers: Kris Manos, Kent Gawart, and Paul Murray (Herman Miller), interview by author, Zeeland, MI, 4 May 2004.

p. 222, thousands have some responsibility: Phil Berry, interview by author, Toronto, Canada, 10 March 2006.

p. 223–224, "no incentive will save your job": Reed, interview.

p. 224, starts all his own staff reviews: Dick Hunter (Dell), interview by author, Parmer, TX, 16 October 2004.

p. 224, some managers didn't seem to buy in: Bergmark, interview.

p. 225, Starbucks Mission Review: Ben Packard and Kevin Carothers (Starbucks), conversation with author, 24 October 2004, and correspondence with author, 22 April 2005.

p. 226, "company that's as good as its word": Bill Nelson and Rick Renner (3M), conversations with author, 9 June 2004.

p. 226, annual environmental report: For more information on reporting, see SustainAbility, United Nations Environment Programme, and Standard & Poor's *Risk and Opportunity: Best Practice in Non-Financial Reporting*, at www.sustainability.com/insight/reporting-article.asp?id=128; Allen White, "New Wine, New Bottles: The Rise of Non-Financial Reporting," *A Business Brief by Business for Social Responsibility*, 20 June 2005, at www.bsr.org/Meta/200506_BSR_Allen-White_Essay.pdf; and Global Reporting Initiative, *Sustainability Reporting Guidelines of 2002*, at www.globalreporting.org/guidelines/2002/contents.asp.

p. 226–227, Northeast Utilities issued and environmental report: "Connecticut Utility Pleads Guilty," *WaterTech online*, 30 September 1999, www.waternet.com/News.asp?mode=4&N_ID=12648. The guilty pleas led to fines and penalties of $10 million.

p. 227, The Global Reporting Initiative: For more on GRI's efforts, see www.globalreporting.org. We've posted some examples of award winning corporate reports at www.eco-advantage.com.

p. 228, "seven years before I talked": Ray Anderson (Interface), interview by author, New Haven, CT, 6 March 2005.

p. 228, "Wall Street will never accept": Gary Pfeiffer, "Keynote," Speech, Conference Board annual meeting on Business and Sustainability, New York, NY, 10 June 2004.

p. 229, "manage the underlying performance": Mark Weintraub and Lex Holst (Shell), interview by author, The Hague, Netherlands, 20 October 2004.

p. 230, "The $6 million savings for Amanco is a magic wand": Maria Emilia Correa, conversation with author, 1 June 2004.

p. 231, EHS First: Dan Gagnier (Alcan), interview with author, Vancouver, Canada, 31 March 2004; Simon Laddychuk and Paola Kistler (Alcan), conversation with author, 23 June 2004.

p. 231, Kaelin planned a demonstration: Kaelin, interview.

p. 231–232, "Notes from your Company Ecologist": Elysa Hammond, interview by author, New Rochelle, NY, 10 December 2004; also correspondence with author 1 March 2006.

p. 232, "it's about a mindset": Laddychuk and Kistler, interview.

PART FOUR. PUTTING IT ALL TOGETHER

CHAPTER 10. WHY ENVIRONMENTAL INITIATIVES FAIL

p. 237, Ford's factory on the Rouge River: The story of the Rouge River plant has been reported in many places. See www.ford.com for more information; also, "Ford Transforms Old Facility into Enviro Friendly Plant," *Greenwire*, 4 April 2004.

p. 240, preserve 138,000 acres: See "Wal-Mart Pledges One Acre for Every Acre Developed," Press Release, 12 April 2005, at http://www.walmartfacts.com/docs/april2005_acres_for_america.pdf.

p. 240–241, "Cod's own country": Dierk Peters, conversation with author, 2 November 2004.

p. 242, Prius sells for $5,000 more: Bloomberg News, "Toyota Says It Plans Eventually To Offer an All-Hybrid Fleet," 14 September 2005.

p. 243–244, coffee-cup sleeve: Ben Packard and Kevin Carothers (Starbucks), conversation with author, 24 October 2004; also Sue Mecklenburg, interview by author, 22 April 2005.

p. 244, "a tension between business performance and environmental goals," interview by author, London, England, 21 October 2004; also correspondence with author, 13 February 2006.

p. 246, PFCs are up to 10,000 times more powerful: In technical terms, PFCs have a very high "global warming potential."

p. 246–247, disposable cameras: Bloomberg News, "Fuji Photo Wins Dispute Over Cameras," *New York Times*, 3 August 2004.

p. 247, environmental professionals now work side-by-side: Lew Scarpace, Tim Mohin, Terry McManus, and Randy Helgeson (Intel), interview by author, Chandler, AZ, 28 January 2004; and Mohin, interview by author, Washington, D.C., 3 November 2005. See also John Harland and Tim Mohin. "Designing for the environment Turns Intel Fabs Green," *Technology@Intel*, November 2005.

p. 248, a real champion: Larry Rogero (Kinko's), conversations with author, 25 November 2003 and 12 December 2003.

p. 250, "six flags on a hill": Larry Rogero (Kinko's), conversation with author, 31 March 2005.

p. 251, wasps in the parking lot: Kris Manos (Herman Miller), interview by author, New Haven, CT, 30 November 2004; also Mark Schurman (Herman Miller), correspondence with author, 28 February 2006.

p. 253–254, three new options for its McNuggets package: Bob Langert, Else Krueck (McDonald's) and Ken Krause (Perseco), interview by author, Oak Brook, IL, 28 July 2004.

p. 256, After the 1992 Earth Summit: Charles O. Holliday, Jr., Stephan Schmidheiny, and Philip Watts, *Walking the Talk: The Business Case for Sustainable Development* (Sheffield, UK: Greenleaf, 2002), 139; discussed in Monica Araya, "To Tell or Not to Tell? Determinants of Environmental Disclosure and Reporting in Corporate Latin America," (Ph.D. Dissertation, Yale University, 2006), 276.

p. 257, "local community reaction was negative": Dave Richards (Rio Tinto), conversation with author, 5 January 2005. See also D. G. Richards, "Integrating Mineral Development and Biodiversity Conservation into Regional Land-Use Planning," *Landscape Ecology and Wildlife Habitat Evaluation: Critical Information for Ecological Risk Assessment, Land-Use Management Activities, and Biodiversity Enhancement Practices*, Eds. L. A. Kapustka, H. Galbraith, M. Luxon, and G. R. Biddinger (West Conshohocken, PA: ASTM International, 2004), ASTM STP 1458.

p. 257, "preferred supplier" program: Sue Mecklenburg (Starbucks), interview by author, 27 March 2005.

p. 258, "They came to our industry": Terry Kellogg (Timberland), conversations with author, 3 December 2003, 25 March 2005, and 8 April 2005.

p. 258–259, SAPREF crude oil refinery: Shell has since invested $49 million in process changes and has cut the refinery's sulfur dioxide emissions by 50 percent over the last five years.

p. 259, "The damage to public opinion within France": "TotalFinaElf: 'We Have Learnt by Experience'," *Weekly Petroleum Argus*, 27 May 2002; also "Damage Limitation," *Weekly Petroleum Argus*, 27 May 2002.

CHAPTER 11. TAKING ACTION

p. 265, DuPont made money: Forest Reinhardt, "Du Pont Freon Products Division," Harvard Business School Case 389111, 1989.

p. 265, Champion paper supported: Forest Reinhardt, "Champion International Corp.: Timber, Trade, and the Northern Spotted Owl," Harvard Business School Case 792017, 1991.

p. 270, Citigroup and Environmental Defense: Tom Murray (Environmental Defense), conversation with author, 3 May 2004; also Tom Murray, "Copy This! Reducing the Environmental Impacts of Copy Paper Use," presentation document, 19 November 2003, provided to author.

p. 271, experimenting with low-impact farming: Jeroen Bordewijk (Unilever), interview by author, Rotterdam, Netherlands, 19 October 2004; Chris Pomfret (Unilever), conversations with author, 19 October 2004 and 1 November 2004.

p. 275, Rio Tinto . . . planning sessions on business drivers: Dave Richards (Rio Tinto), conversation with author, 5 January 2005.

p. 276, seemingly secondary issues: Pat Nathan (Dell), interview by author, Round Rock, TX, 16 October 2004; Tod Arbogast (Dell), correspondence with author, 24 February 2006.

p. 278, "eWheel": Goran Brohammer (IKEA), interview by author, Almhult, Sweden, 24 August 2004.

p. 279–281, a more complete stakeholder strategy: The topic of how stakeholders fit into the theory of the firm has been a focus of academic research for many years. Most researchers cite the following book as the starting point for modern thinking on the subject: R E. Freeman, *Strategic Management: A Stakeholder Approach* (Boston: Pitman, 1984). We drew in particular on two analyses that lay out best practices and frameworks for prioritizing stakeholders: Ronald K. Mitchell, Bradley R. Agle, and Donna J. Wood, "Toward a Theory of Stakeholder Identification and Salience: Defining the Principle of Who and What Really Counts," *Academy of Management Review*, October 1997, 22, 4; and Jem Bendell, "Talking for Change? Reflections on Effective Stakeholder Dialogue," New Academy of Business, 20 October 2000.

p. 280, rating stakeholders . . . on three dimensions: Mitchell et al., "Toward a Theory of Stakeholder Identification and Salience."

p. 281, maps stakeholders based on . . . : For more on this two-by-two matrix, see G. Savage, T. Nix, C. Whitehead, and J. Blair, "Strategies for Assessing and Managing Organizational Stakeholders," *The Executive* 5, 1991, no. 2, 61–75.

CHAPTER 12. ECO-ADVANTAGE STRATEGY

p. 283, Michael Porter's highly regarded strategy model: Michael Porter, *Competitive Advantage: Creating and Sustaining Superior Performance* (New York: Free Press, 1985).

p. 284, executives who possess the ability for integrated thinking: See Howard Gardner, "The Synthesizing Leader" in "The HBR List: Breakthrough Ideas for 2006," *Harvard Business Review,* February 2006.

p. 288, After a CEO pow-wow on climate change: Adam Aston and Burt Helm, "The Race Against Climate Change," *BusinessWeek*, 12 December 2005, 59.

p. 291, Romanoff changed one of its products: "Home Depot Pushes Contractor to Use Environment-Friendly Product," *Greenwire* from GreenBiz.com, 7 June 2003.

p. 291–292, chart borrowers on a classic two-by-two matrix: André Abadie, interview by author, New York, NY, 2 June 2005.

p. 298, the company saved $173 million: Aston and Helm, "The Race Against Climate Change," p. 66.

p. 299, cut rejects from the Lycra production line: Paul Tebo, Dawn Rittenhouse, and Ed Mongan (DuPont), interview by author, Wilmington, DE, 5 March 2004. This line of business is now part of Invista, owned by Koch Industries.

p. 299, ship its hazardous waste: Alex Heard (Intel), conversation with author, 29 January 2004.

p. 300, "try for a market benefit": Chris Mottershead (BP), interview by author, London, England, 21 October 2004.

p. 300, sales in this market leapt 44 percent: Mark Stanland (Bay West), conversations with author, 8 September 2004 and 3 April 2006; Arthur Weissman (Green Seal), conversation with author, 18 May 2004.

p. 302, the Hat Trick Green-to-Gold Play at Alcan: Travis Engen, "Keynote," speech, Globe 2004, Vancouver, BC, Canada, 31 March 2004; Dan Gagnier (Alcan), interview with author, Vancouver, Canada, 31 March 2004; Simon Laddychuk and Paola Kistler (Alcan), conversation with author, 23 June 2004.

p. 304, "That's not limousine liberal": Unmesh Kher, "Getting Smart at Being Good . . . Are Companies Better Off for It?" *TIME Inside Business, TIME Magazine,* January 2006, A7.

Index

Note: Page numbers in italic reflect text in graphic elements.